AF352776

Engineering Fluid Dynamics 2019–2020

Engineering Fluid Dynamics 2019–2020

Volume 2

Editor

Bjørn H. Hjertager

MDPI • Basel • Beijing • Wuhan • Barcelona • Belgrade • Manchester • Tokyo • Cluj • Tianjin

Editor
Bjørn H. Hjertager
University of Stavanger
Norway

Editorial Office
MDPI
St. Alban-Anlage 66
4052 Basel, Switzerland

This is a reprint of articles from the Special Issue published online in the open access journal *Energies* (ISSN 1996-1073) (available at: https://www.mdpi.com/journal/energies/special_issues/eng_fluid_dyn2019-2020).

For citation purposes, cite each article independently as indicated on the article page online and as indicated below:

LastName, A.A.; LastName, B.B.; LastName, C.C. Article Title. *Journal Name* **Year**, *Volume Number*, Page Range.

Volume 2
ISBN 978-3-0365-0250-2 (Hbk)
ISBN 978-3-0365-0251-9 (PDF)

Volume 1-2
ISBN 978-3-0365-0252-6 (Hbk)
ISBN 978-3-0365-0253-3 (PDF)

Contents

About the Editor

Bjørn H. Hjertager received his Ph.D. from the University of Trondheim, Norway (now the Norwegian University of Science and Technology, NTNU), in 1979. His thesis examined combustion, heat transfer, and fluid flow. Thereafter, he stayed for almost 10 years at Chr. Michelsen Institute. From the late 1980s to the late 1990s, he worked at the Telemark Institute of Technology (HiT-TF), and at the research institute Tel-Tek. He then spent 11 years at Aalborg University in Denmark. Since 2008, he has returned to Norway, as Professor in Fluid Dynamics at the University of Stavanger. He has published more than 180 papers on fluid flow, heat transfer, combustion, gas explosions, and chemical reactors, and has supervised 21 Ph.D. candidates in Norway and Denmark.

Preface to "Engineering Fluid Dynamics 2019–2020"

This book contains the successful submissions to a Special Issue of Energies entitled"Engineering Fluid Dynamics 2019–2020". The topic of engineering fluid dynamics includes both experimental and computational studies. Of special interest were submissions from the fields of mechanical, chemical, marine, safety, and energy engineering. We welcomed original research articles and review articles. After one-and-a-half years, 59 papers were submitted and 31 were accepted for publication. The average processing time was about 41 days. The authors had the following geographical distribution: China (15); Korea (7); Japan (3); Norway (2); Sweden (2); Vietnam (2); Australia (1); Denmark (1); Germany (1); Mexico (1); Poland (1); Saudi Arabia (1); USA (1); Serbia (1). Papers covered a wide range of topics including analysis of free-surface waves, bridge girders, gear boxes, hills, radiation heat transfer, spillways, turbulent flames, pipe flow, open channels, jets, combustion chambers, welding, sprinkler, slug flow, turbines, thermoelectric power generation, airfoils, bed formation, fires in tunnels, shell-and-tube heat exchangers, and pumps. I found the task of editing and selecting papers for this collection to be both stimulating and rewarding, and I would like to thank the staff and reviewers for their efforts and input.

Bjørn H. Hjertager
Editor

Energy Efficiency of Pneumatic Cylinder Control with Different Levels of Compressed Air Pressure and Clamping Cartridge

Vladislav Blagojevic [1,*], Dragan Seslija [2], Slobodan Dudic [2] and Sasa Randjelovic [1]

[1]   Faculty of Mechanical Engineering, University of Nis, Aleksandra Medvedeva 14, 18000 Nis, Serbia; sasa.randjelovic@masfak.ni.ac.rs

[2]   Faculty of Technical Sciences, University of Novi Sad, Trg Dositeja Obradovica 6, 21000 Novi Sad, Serbia; seslija@uns.ac.rs (D.S.); dudic@uns.ac.rs (S.D.)

*   Correspondence: vladislav.blagojevic@masfak.ni.ac.rs

Received: 27 June 2020; Accepted: 15 July 2020; Published: 19 July 2020

Abstract: Since pneumatic systems are widely used in various branches of industry, the need to find ways to reduce energy consumption in these systems has become very pressing. The reduction in energy consumption in these systems is reflected in the reduction of compressed air consumption. The paper presents a cylinder control system with a piston rod on one side, in which the reduction in energy consumption is ensured by using different levels of supply pressure in the working and the return stroke, and by holding the cylinder piston rod in its final positions with a clamping cartridge. Clamping and holding the piston rod in its final position further affects the reduction in energy consumption. Experimental data show that the application of the proposed control leads to a decrease in compressed air consumption of 25.54% to 32.97%, depending on the compressed air pressure used in the return stroke. The cost-effectiveness of the proposed cylinder control with different levels of compressed air pressure and holding the final position by clamping cartridge is presented.

Keywords: control; cylinder; energy efficiency; clamping; pneumatics

---

1. Introduction

Together with pneumatic actuators, pneumatic systems are widely used in numerous industrial applications. These systems use compressed air in their operation, such air being usually produced by electrical energy that drives a compressor. The electricity costs for compressed air production can constitute up to 20% of the entire electricity costs in an industry [1]. This is one of the most important reasons behind the search for ways to save energy when using pneumatic systems.

Numerous authors have dealt with the problem of energy efficiency of pneumatic systems and especially reduction of compressed air consumption. Al-Dakkan et al. [2] presented a method of energy saving in the context of a servo-controlled pneumatic actuator. Yang et al. [3,4] proposed a new booster valve with energy recovery for improving the energy efficiency of a pneumatic actuator system. Luciano et al. [5] introduced an alternative scheme with a fast switching on/off valve interconnecting the cylinder chambers, for reducing the consumption of compressed air in pneumatic positioning systems with external loads. Joshua et al. [6] introduced a pneumatic strain energy accumulator for recycling exhaust air from one pneumatic component, storing it in a highly efficient process and reusing the stored exhaust air at a constant pressure to power another pneumatic component. Wang et al. [7] used exergy-related analysis for evaluating the efficiency of pneumatic systems. Shen et al. [8] and Yang et al. [9] proposed an energy saving approach by supplementing a standard spool valve-controlled pneumatic actuator with an additional two-way valve that enables flow between the cylinder chambers. Seslija et al. [10] introduced the application of pulse-width modulation and by-pass chamber control of the pneumatic rodless cylinder. Bartyś et al. [11] proposed three practical measures of electro-pneumatic control quality, namely: variability, mean time, and cumulative effort. These measures are very useful in enabling optimization of positioner controller settings with respect to the controller effort and improving the energy efficiency of pneumatic systems. Kanno et al. [12] developed a three-port poppet-type servo valve to reduce air leakage and to improve the energy efficiency of the whole pneumatic system. Many authors wereimproving the energy efficiency in the driven actuator and making the circuit more complex, such as using dual pressure supply, utilizing expansion energy [13], recovering energy with a rubber bladder and storing the strain energy [14], or reusing exhaust air for power generation [15]. In addition, reusing exhaust air as input to the drive chamber of the cylinder led to velocity fluctuations [16]. Nehler T. [17] reviewed the existing base of scientific knowledge on energy efficiency in compressed air systems and suggested that energy efficiency measures in compressed air systems and related non-energy benefits should be studied on a specific measure level to fully understand and acknowledge their effects on the energy use of a compressed air system and possible additional effects, i.e., non-energy benefits.

All of the previous papers dealt with the problem of increasing the energy efficiency of pneumatic systems in various ways. Some of them proposed new booster valves [3,4]. Others reused exhaust air from one pneumatic component to the same component [5,8–10,15,16] or another component [6,15]. There were papers that proposed a new algorithm of control [2], optimization of positioner controller settings [11], and reducing air leakage to improve the energy efficiency of the whole pneumatic system [12].

This paper deals with the problem of decreasing compressed air consumption in executive pneumatic systems in a new way. The goal of this paper is to develop a new pneumatic system that uses various levels of compressed air in the working and the return cylinder stroke as well as keeps the cylinder piston rod at rest in final positions by a clamping cartridge, during which the supply to the cylinder chambers is cut off. The proposed pneumatic system, with various levels of compressed air supply and clamping cartridge, enables the reduction of compressed air consumption and at the same time an increase in the energy efficiency of the whole pneumatic system. If a lower compressed air pressure is used to supply the cylinder chambers, the compressed air consumption will be reduced. The proposed cylinder control system is presented in Section 2. In addition to the fact that using lower compressed air in the return cylinder stroke enables the reduction of compressed air consumption, it also leads to a decrease in the velocity of cylinder piston movement, which is explained and discussed in Sections 3 and 4. In Section 4.1, the cost-effectiveness of the proposed system is discussed.

2. Cylinder Control System with Different Levels of Compressed Air Pressure and Holding the Final Position by Clamping Cartridge

The most often used executive part in conventional pneumatic systems comprises various double-acting pneumatic cylinders and a supply valve, and the same level of compressed air pressure Ps is used both in the working and in the return stroke of the cylinder. Figure 1 shows a typical pneumatic executive system with an actuator in the form of a double-acting cylinder with a piston rod on one side marked 1.0, and a 5/2-way monostable valve with electrical activation signal y marked 1.1.

A pneumatic executive control system with different levels of compressed air pressure and holding the final position (Figure 1b) enables a reduction in compressed air consumption.

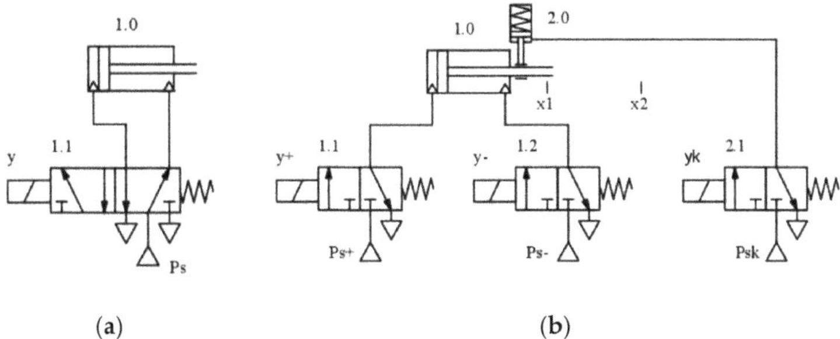

Figure 1. A pneumatic executive system: (**a**) typical; (**b**) with different levels of compressed air pressure and holding the cylinder piston rod in its final position.

In the return stroke of the cylinder, it is possible to carry out the process using a lower level of air pressure and reduce consumption of compressed air. A pneumatic executive system that enables the use of different levels of compressed air pressure in the working and the return stroke of the cylinder requires the substitution of a common supply valve with two valves that allow for an independent supply of different levels of compressed air pressure to the cylinder chambers. In the present case, this is done by substituting a 5/2-way valve marked 1.1 in Figure 1a with two 3/2-way valves marked 1.1 and 1.2 in Figure 1b. The y+ and y- signals are the signals of the activation of valves 1.1 and 1.2, respectively, which serve to supply the cylinder chambers with an appropriate compressed air pressure. The simultaneous activation of y+ and y- signals is forbidden and it is called "prohibited state". In Figure 1b, when the cylinder piston moves right, which represents the working stroke, valve 1.1 is activated and supplies the left cylinder chamber with a higher-pressure P_s+. In the return stroke, valve 1.2 is activated and the return stroke is performed, i.e., the cylinder piston moves to the left at a lower pressure P_s−. It is a well-known fact that as a cylinder moves, the level of pressure in its chamber falls below the level of supply, and the pressure rises to the maximum only at the end of the stroke, when the cylinder reaches its final position. This fact can be used to further reduce compressed air consumption by stopping the cylinder chamber's supply at the end of the stroke and keeping the cylinder in its final position. This will prevent the rise of pressure in the cylinder chambers to its maximum. To do this, it is necessary to add cylinder clamping cartridges to the pneumatic executive system. The clamping cartridge 2.0 is executed in such a way that it releases the cylinder 1.0 only when compressed air is supplied to it, Figure 1b. To supply this clamping cartridge, the 3/2-way valve marked 2.1 is used, Figure 1b. The level of compressed air P_{sk} for the supply of the clamping cartridge is equal to the level of the pressure necessary for the working stroke of the P_s+ cylinder. This is done to ensure that the cylinder is kept in its final position in a proper and safe manner. The detection of final positions of the cylinder is enabled by the limited sensors x_1 for the drawn-in and x_2 for the drawn-out position. In the working stroke, the cylinder is supplied with the P_s+ pressure, while the lower P_s− pressure is used in the return stroke.

A Mathematical Model of Cylinder Control with Different Levels of Compressed Air Pressure and Holding the Final Position by Clamping Cartridge

The mathematical model of the cylinder control system with different levels of compressed air pressure and holding the final position consists of the dynamic model of the cylinder, the model of pressure change in the cylinder chambers, and the model of flow change in the supply valve.

The dynamic model of the cylinder is Equation (1),

$$(M_L + M_P)\ddot{x} + \beta\dot{x} + F_f + F_L = P_1 A_1 - P_2 A_2, \tag{1}$$

where M_L is the mass of the external parts connected to the cylinder, M_P the mass of the cylinder piston and the piston rod, x the position of the piston, β the coefficient of viscous friction, F_f the Coulomb friction force, F_L the external force, P_1 and P_2 the absolute pressures in the cylinder chambers with their maximums being P_s+ or P_s-, depending on the level of compressed air used to supply the cylinder chambers, and A_1 and A_2 the effective surface areas of the piston and the piston rod chamber of the cylinder.

The righthand side of Equation (1) represents the cylinder active force, produced by the pressure differential acting across the cylinder piston. The existence of this force results in the displacement of the cylinder piston and the piston rod to the one or the other side. This active force directly influences the velocity of cylinder piston movement. If the lower compressed air pressure is used to supply the cylinder chambers, the active force will be lower, and the cylinder piston movement will be slower.

The equation that describes the model of the pressure change in the pneumatic cylinder chambers is Equation (2),

$$\dot{P}_i = \frac{RT}{V_{0i} + A_i\left(\frac{1}{2}L \pm x\right)}\left(\alpha_{in}\dot{m}_{in} - \alpha_{out}\dot{m}_{out}\right) \mp \alpha\frac{P_i A_i}{V_{0i} + A_i\left(\frac{1}{2}L \pm x\right)}\dot{x}, \; i = 1,2, \tag{2}$$

where R is the ideal gas constant, T the temperature, $\dot{m}_{in}$ and $\dot{m}_{out}$ the intake and exhaust mass flows from the cylinder chamber, respectively, V_{0i} the inactive cylinder volumes, L the cylinder stroke length, and α α_{in}, α_{out}, the thermal coefficients. For the process of filling the chambers with compressed air, $\alpha_{in} = \kappa = 1.4$, while $\alpha_{out} = 1$ for the emptying process [17]. The thermal characteristics of the processes taking place while the piston is moving are better described if one understands that $\alpha = 1.2$ [17]. Due to simplicity, the heat transfer losses have not been considered for the calculation of the gas temperature.

Because of simplicity, it is assumed that the length of the lines between the cylinder and the 3/2-way supply valves is not long, so that the intake and exhaust mass flows from the cylinder chambers depend only on the flow through the valve openings, as presented in Equation (3),

$$\dot{m}_v = \begin{cases} c_f A_v C_1 \dfrac{P_u}{\sqrt{T}} & za \; \dfrac{P_d}{P_u} \le p_{cr}, \\[2em] c_f A_v C_2 \dfrac{P_u}{\sqrt{T}}\left(\dfrac{P_d}{P_u}\right)^{\frac{1}{\kappa}}\sqrt{1 - \left(\dfrac{P_d}{P_u}\right)^{\frac{\kappa-1}{\kappa}}} & za \; \dfrac{P_d}{P_u} > p_{cr}, \end{cases}, \; v = in, out, \tag{3}$$

where c_f is the non-dimensional emptying coefficient, and P_u and P_d the upstream and the downstream pressure, respectively. The value A_v represents the surface area of the 3/2-way supply valve opening, and it can have the value of $A_v = A_v\text{sgn}(U_v)$, where U_v is the voltage of the 3/2-way valve activation.

The values C_1, C_2 and P_{cr}, in Equation (3), are constants that depend on the fluid, in this case air ($\kappa = 1.4$) [17],

$$C_1 = \sqrt{\frac{\kappa}{R}\left(\frac{2}{\kappa+1}\right)^{\frac{\kappa+1}{\kappa-1}}} = 0.040418, \; C_2 = \sqrt{\frac{2\kappa}{R(\kappa-1)}} = 0.156174,$$
$$p_{cr} = \left(\frac{2}{\kappa+1}\right)^{\frac{\kappa}{\kappa-1}} = 0.528. \tag{4}$$

The unit that holds the cylinder in its final position also consumes a certain amount of compressed air. Since it represents a single-acting cylinder with a spring-operated return stroke (Figure 1b), the stroke and the dimensions of its work chamber are small; thus, the air consumption is not significant, because the chamber is supplied with the appropriate amount of compressed air very quickly, releasing the cylinder. The volume of the by clamping cartridge work chamber is much smaller than the volumes of the work chambers of the cylinder itself. The cylinder is held again by simply cutting off the supply of the by clamping cartridge.

Equation (3) shows that the mass flow of compressed air and its consumption are influenced, among other things, by the upper pressure P_u that can be maximally equal to P_s+. If a lower supply pressure is used as P_u, $P_s- < P_s+$, both the mass flow and the consumption decrease, and it also leads to a decrease in the velocity of cylinder piston movement.

The value of the lower supply pressure must be sufficient to exert the appropriate force in the cylinder to return it to the initial position. The asterisk in Figure 2 shows the reduction in the level of compressed air flow when different levels of supply pressure are used, which directly impacts the increase in energy efficiency.

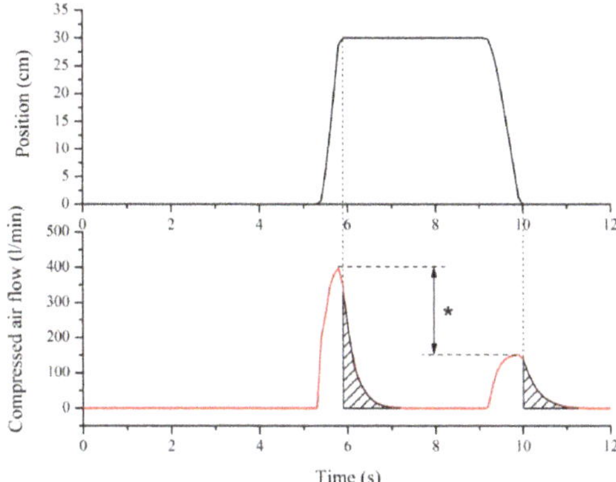

Figure 2. Compressed air flow when a supply pressure of 600 kPa is used in the working stroke and 300 kPa in the return stroke.

Equation (2) provides the conclusion that the pressure in the cylinder chambers increases up to the maximum value, i.e., the value of the supply pressure, only after reaching the final position. If the supply is cut off at the end of the cylinder stroke and the cylinder is held in its final position, this directly affects the reduction in air consumption, since the pressure in the work chamber is prevented from reaching its maximal value, as shown by the hatched area in Figure 2.

3. Results

To present the advantage of using the cylinder control system with different levels of compressed air pressure in the working and the return stroke, as well as the advantage of holding the cylinder piston in its final position, experiments were performed with universal cylinders with a piston rod on one side, manufactured by FESTO, with different piston diameters of 32, 40, 63, 80, 100 and 125 mm, and a 250 mm stroke. For each of the cylinders, 10 measurements were performed. This was done to measure the savings in compressed air consumption by using the proposed cylinder control with different chamber volumes. Two sets of measurements were performed.

Firstly, compressed air consumption measurements were performed at the supply pressure of 600 kPa, on cylinders supplied by a 5/2-way valve, Figure 1a. This is a common cylinder—supply valve link. The average air consumption results, with relative error 1.2%, are shown in Table 1.

Table 1. Air consumption results for the system in Figure 1a.

Piston Diameter (mm)	Compressed Air Consumption (l)
32	3.58
40	4.64
63	10.65
80	18.86
100	29.48
125	45.77

Based on these compressed air consumption results, the savings during other measurements were calculated.

The second set of measurements was performed with the supply pressure P_s+ set at 600 kPa in the working stroke and the lower pressures P_s- at 500, 400 and 300 kPa in the return stroke of the cylinder, Figure 1b. The supply pressure of the cylinder with the clamping cartridge was equal to the supply pressure of the cylinder in the working stroke, i.e., $P_{sk} = P_s+ = 600$ kPa. Compressed air consumption of the by clamping cartridge depended on the diameter of the used cylinder, i.e., the diameter of the piston rod that was supposed to be held, and it ranged from 0.02 L for the 32 mm piston diameter, up to 0.9 L for the 125 mm piston diameter. The specific components used in these measurements are shown in Figure 1b.

The average air consumption results, with relative error 1.2%, are shown in Table 2.

Table 2. Air consumption results for the system in Figure 1b.

Piston Diameter (mm)	Compressed Air Consumption with Return Stroke Pressure of 500 kPa (l)	Compressed Air Consumption with Return Stroke Pressure of 400 kPa (l)	Compressed Air Consumption with Return Stroke Pressure of 300 kPa (l)
32	2.64	2.52	2.4
40	3.43	3.27	3.11
63	7.87	7.47	7.14
80	14.03	13.48	12.76
100	21.95	20.82	19.91
125	33.92	32.45	30.9

As was explained in the previous section, if the lower compressed air pressure is used to supply the cylinder chambers, the active force will be lower, and the cylinder piston movement will be slower. The diagrams in Figures 3–5 show the experimental results of dependences of velocity of cylinder piston movement on compressed air pressure in the return stroke.

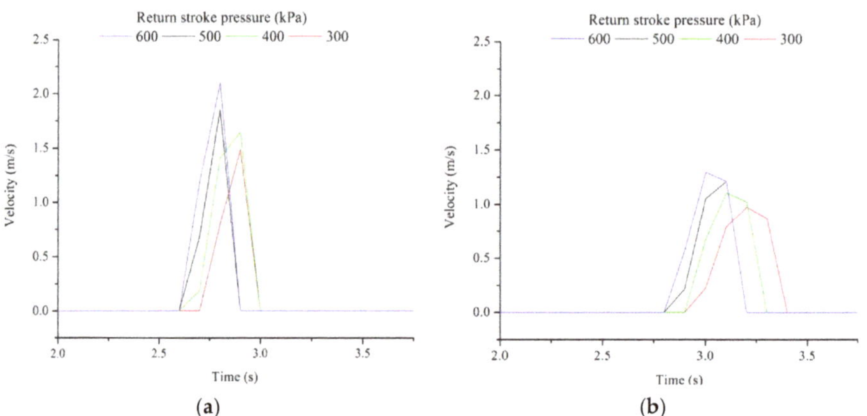

Figure 3. Velocity of cylinder piston movement. Diameter of cylinder: (**a**) 32 mm; (**b**) 40 mm.

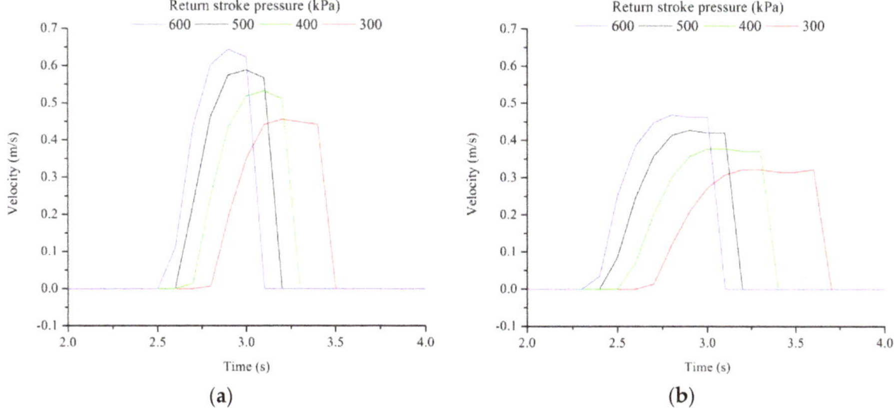

Figure 4. Velocity of cylinder piston movement. Diameter of cylinder: (**a**) 63 mm; (**b**) 80 mm.

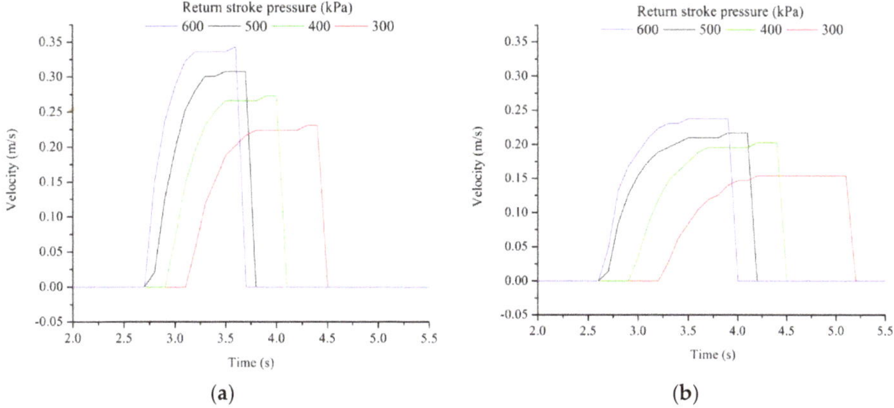

Figure 5. Velocity of cylinder piston movement. Diameter of cylinder: (**a**) 100 mm; (**b**) 125 mm.

4. Discussion

Based on the obtained results of the compressed air consumption measurements from Tables 1 and 2, a reduction in compressed air consumption can be observed when lower supply pressure is used in the return stroke and when the cylinder is held by a clamping cartridge in accordance with Figure 1b.

Table 3 shows the possible compressed air savings that could be obtained by using different levels of supply pressures in the working and the return stroke, when the unit for holding the cylinder in its final position is used. From Table 3, it can observed that the average savings in compressed air consumption when lower supply pressures are used in the return stroke, and with the use of a clamping cartridge for holding the cylinder in its final positions, are as follows: 25.91% when $P_s- = 500$ kPa, 29.33% when $P_s- = 400$ kPa, and 32.7% when $P_s- = 300$ kPa.

Table 3. The savings in compressed air consumption after second set of measurements.

Piston Diameter (mm)	Compressed Air Consumption with Return Stroke Pressure of 500 kPa (l)	Compressed Air Consumption with Return Stroke Pressure of 400 kPa (l)	Compressed Air Consumption with Return Stroke Pressure of 300 kPa (l)
32	26.26	29.61	32.96
40	26.08	29.53	32.97
63	26.1	29.86	32.96
80	25.61	28.53	32.34
100	25.54	29.38	32.46
125	25.89	29.1	32.49

Based on the above, and from the perspective of energy savings, it can be concluded that the lower the supply pressure used in the return stroke, the better. This claim has its limitations in the sense that when lower pressures are applied in the return stroke, the stroke itself, i.e., the return of the cylinder to its initial position, slows down, meaning that more time is needed for it to be fully performed. The velocity of cylinder piston movement in the return stroke directly depends on compressed air pressure, Figures 3–5. If the velocity of cylinder piston movement is slower, then the duration of movement is longer. The duration of cylinder piston movement in the return stroke, depending on compressed air pressure, is shown in Table 4. The pressure in the return stroke can be lowered, but it needs to remain sufficiently high to ensure the movement of the cylinder.

Table 4. The duration of cylinder piston movement in the return stroke.

Piston Diameter (mm)	Duration of Movement with Return Stroke Pressure of 600 kPa (s)	Duration of Movement with Return Stroke Pressure of 500 kPa (s)	Duration of Movement with Return Stroke Pressure of 400 kPa (s)	Duration of Movement with Return Stroke Pressure of 300 kPa (s)
32	0.3	0.3	0.4	0.4
40	0.4	0.4	0.5	0.6
63	0.6	0.7	0.8	1
80	0.8	0.9	1.1	1.4
100	1	1.1	1.4	1.8
125	1.4	1.6	1.9	2.6

4.1. Cost-Effectiveness

In order to apply the cylinder control system with different levels of compressed air pressure and holding the final position, it is necessary to replace the 5/2-way monostable valve, Figure 1a, with three 3/2-way monostable valves and to install a clamping cartridge, Figure 1b.

The cost-effectiveness of the proposed control is based on the price of FESTO company products for clamping cartridges with designation KP, AZ PNEUMATICA company products for valves, and a cost of compressed air of 0.022 €/m^3. The return of investment period (RIP) depends on the saving (%) accomplished by the proposed control and the number of cylinder working cycles per year, or cycle duration. If the cycle duration is shorter, the number of cylinder working cycles per year is higher, when the cylinder works all the time. According to previous assumptions, the number of cylinder working cycles per year (NCWCY) is NCWCY = (52 weeks × 5 days × 2 shifts × 8 h × 60 min × 60 s)/*tc*, where *tc* is cycle duration in seconds. Prices of additional components for holding the cylinder in final positions range from €357.60 to €1746.80, and prices for valves range from €26.50 to €46.90, depending on cylinder dimensions.

The diagrams in Figures 6–8 show the RIP, depending on the cycle duration and return stroke pressure of the cylinder.

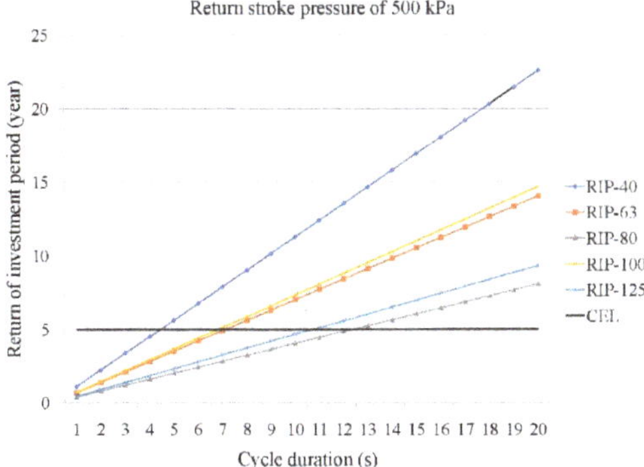

Figure 6. Return of investment period (RIP) with return stroke pressure of 500 kPa.

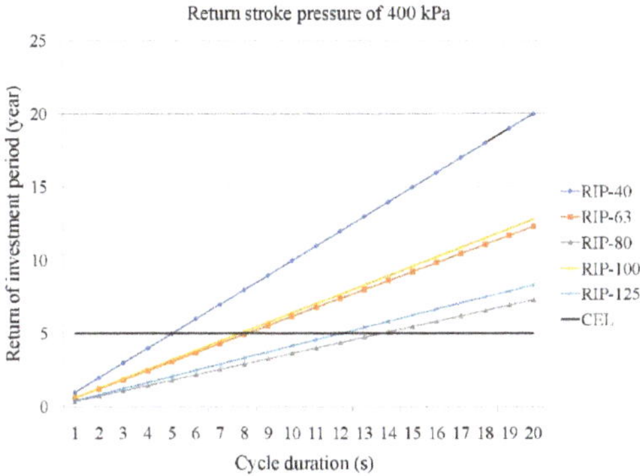

Figure 7. Return of investment period (RIP) with return stroke pressure of 400 kPa.

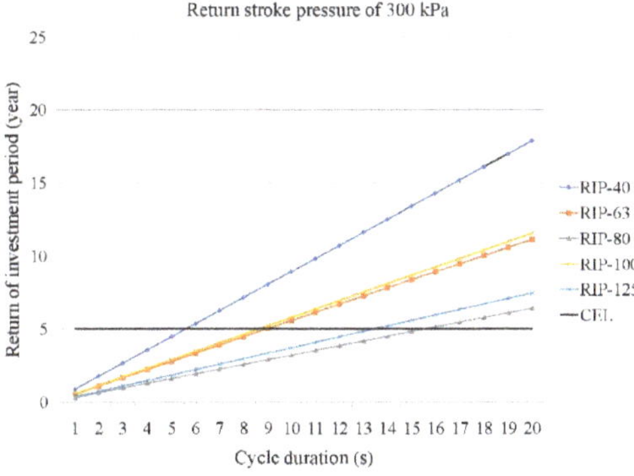

Figure 8. Return of investment period (RIP) with return stroke pressure of 300 kPa.

The cost-effectiveness limit (CEL) of five years shows the value of cost-effective cycle duration for different cylinders and return stroke pressures. If the RIP is less than CEL, the cycle duration is cost-effective for implementing the proposed cylinder control with clamping cartridge and different levels of compressed air pressure. For pneumatic cylinders with larger chamber volume, cost-effective cycle duration is even longer.

5. Conclusions

The control system of a double-acting cylinder with a piston rod on one side using different pressure levels of compressed air in the working and the return stroke, and with the clamping cartridge for holding the cylinder piston rod in its final positions, shows good characteristics and ensures savings. Savings in compressed air consumption in the proposed system range from 25.54% to 32.97%, depending on the air pressure used in the return stroke. The average savings in compressed air consumption for different cylinders are as follows: 25.91% when $P_s- = 500$ kPa, 29.33% when $P_s- = 400$ kPa, and 32.7% when $P_s- = 300$ kPa. One of the shortcomings of such a system is that a lower cylinder supply pressure in the return stroke leads to a reduced speed of movement of the cylinder piston, which slightly extends the return time and whole cycle duration. With the extension of the cycle duration, the cost-effectiveness of the proposed control decreases, and the RIP becomes higher than five years, regardless of the energy savings in compressed air consumption.

Author Contributions: V.B. and D.S. conceived and designed the experiments; S.D. performed the experiments and analyzed the data; V.B. and S.R. wrote the paper and modified the manuscript. All authors have read and agreed to the published version of the manuscript.

Funding: This research received no external funding.

Conflicts of Interest: The authors declare no conflict of interest.

References

1. Blagojevic, V. Contribution to the Development of Efficient Control of Pneumatic Executive Organs. Ph.D. Thesis, University of Novi Sad, Faculty of Technical Science, Novi Sad, Serbia, 2010.
2. Al-Dakkan, K.A.; Goldfarb, M.; Barth, E.J. Energy saving control for pneumatic servo systems. In Proceedings of the 2003 IEEE/ASME International Conference on Advanced Intelligent Mechatronics (AIM 2003), Kobe, Japan, 20–24 July 2003; pp. 284–289.
3. Yang, F.; Tadano, K.; Li, G.; Kagawa, T.; Peng, J. Simulation on the Characteristics of Pneumatic Booster Valve with Energy Recovery, AsiaSim 2016, SCS AutumnSim 2016. *Commun. Comput. Inf. Sci.* **2016**, *645*, 143–153.
4. Yang, F.; Tadano, K.; Li, G.; Kagawa, T. Analysis of the Energy Efficiency of a Pneumatic Booster Regulator with Energy Recovery. *Appl. Sci.* **2017**, *7*, 816. [CrossRef]
5. Luciano, E.; Victor, J.; De, N.; Eugênio, B.C. Compressed air saving in symmetrical and asymmetrical pneumatic positioning systems. *Proc. Inst. Mech. Eng. Part I* **2015**, *229*, 957–969.
6. Joshua, J.C.; Christopher, J.N.; Seth, T.; Aaron, J.; Sankaran, M.; Eric, J.B. Energy conservation in industrial pneumatics: A state model for predicting energetic savings using a novel pneumatic strain energy accumulator. *Appl. Energy* **2017**, *198*, 239–249.
7. Wang, Z.; Xiong, W.; Wang, H. Exergy analysis of the pneumatic line throwing system. *Int. J. Exergy* **2017**, *19*, 364–379. [CrossRef]
8. Shen, X.; Goldfarb, M. Energy Saving in Pneumatic Servo Control Utilizing Interchamber Cross-Flow. *J. Dyn. Syst. Meas. Control* **2007**, *129*, 303–310. [CrossRef]
9. Yang, A.; Pu, J.; Wong, C.B.; Moore, P. By-pass valve control to improve energy efficiency of pneumatic drive system. *Control Eng. Pract.* **2009**, *17*, 623–628. [CrossRef]
10. Šešlija, D.; Čajetinac, S.; Blagojević, V.; Šulc, J. Application of pulse width modulation and by-pass valve control for increasing energy efficiency of pneumatic actuator system. *Proc. Inst. Mech. Eng. Part I* **2018**, *232*, 1314–1324. [CrossRef]
11. Bartyś, M.; Hryniewicki, B. The Trade-Off between the Controller Effort and Control Quality on Example of an Electro-Pneumatic Final Control Element. *Actuators* **2019**, *8*, 23. [CrossRef]
12. Kanno, T.; Hasegawa, T.; Miyazaki, T.; Yamamoto, N.; Haraguchi, D.; Kawashima, K. Development of a Poppet-Type Pneumatic Servo Valve. *Appl. Sci.* **2018**, *8*, 2094. [CrossRef]
13. Harris, P.; Nolan, S.; O'Donnell, G.E. Energy optimisation of pneumatic actuator systems in manufacturing. *J. Clean. Prod.* **2014**, *72*, 35–45. [CrossRef]

14. Cummins, J.J.; Barth, E.J.; Adams, D.E. Modeling of a pneumatic strain energy accumulator for variable system configurations with quantified projections of energy efficiency increases. In *ASME/BATH 2015 Symposium on Fluid Power and Motion Control*; American Society of Mechanical Engineers: Chicago, IL, USA, 2015; V001T01A055.ASME.

15. Luo, X.; Wang, J.H.; Sun, H.; Derby, J.W.; Mangan, S.J. Study of a new strategy for pneumatic actuator system energy efficiency improvement via the scroll expander technology. *IEEE/ASME Trans. Mechatron.* **2013**, *18*, 1508–1518. [CrossRef]

16. Luo, X.; Sun, H.; Wang, J.H. An energy efficient pneumatic-electrical system and control strategy development. In Proceedings of the 2011 American Control Conference, San Francisco, CA, USA, 29 June–1 July 2011.

17. Nehler, T. Linking energy efficiency measures in industrial compressed air systems with non-energy benefits—A review. *Renew. Sustain. Energy Rev.* **2018**, *89*, 72–87. [CrossRef]

Article

Comparative Study of CFD and LedaFlow Models for Riser-Induced Slug Flow

Rasmus Thy Jørgensen [†], Gunvor Rossen Tonnesen [†], Matthias Mandø and Simon Pedersen [*]

Department of Energy Technology, Aalborg University, Niels Bohrs vej 8, 6700 Esbjerg, Denmark;
rjarge15@student.aau.dk (R.T.J.); gtonne15@student.aau.dk (G.R.T.); mma@et.aau.dk (M.M.)
* Correspondence: spe@et.aau.dk; Tel.: +45-9940-3376
† These authors contributed equally to this work.

Received: 18 June 2020; Accepted: 9 July 2020; Published: 20 July 2020

Abstract: The goal of this study is to compare mainstream Computational Fluid Dynamics (CFD) with the widely used 1D transient model LedaFlow in their ability to predict riser induced slug flow and to determine if it is relevant for the offshore oil and gas industry to consider making the switch from LedaFlow to CFD. Presently, the industry use relatively simple 1D-models, such as LedaFlow, to predict flow patterns in pipelines. The reduction in cost of computational power in recent years have made it relevant to compare the performance of these codes with high fidelity CFD simulations. A laboratory test facility was used to obtain data for pressure and mass flow rates for the two-phase flow of air and water. A benchmark case of slug flow served for evaluation of the numerical models. A 3D unsteady CFD simulation was performed based on Reynolds-Averaged Navier-Stokes (RANS) formulation and the Volume of Fluid (VOF) model using the open-source CFD code OpenFOAM. Unsteady simulations using the commercial 1D LedaFlow solver were performed using the same boundary conditions and fluid properties as the CFD simulation. Both the CFD and LedaFlow model underpredicted the experimentally determined slug frequency by 22% and 16% respectively. Both models predicted a classical blowout, in which the riser is completely evacuated of water, while only a partial evacuation of the riser was observed experimentally. The CFD model had a runtime of 57 h while the LedaFlow model had a runtime of 13 min. It can be concluded that the prediction capabilities of the CFD and LedaFlow models are similar for riser-induced slug flow while the CFD model is much more computational intensive.

Keywords: unsteady RANS simulation; two-phase flow; riser-induced slug flow; LedaFlow; VOF-model

---

1. Introduction

The accurate prediction of the two-phase flow in riser pipeline systems are important for the design and process control of offshore oil and gas facilities [1,2]. The 1D commercial code LedaFlow is commonly used in the industry to predict the undesired slug formation and peak pressure levels [3]. Advances in available computational resources have made it possible to consider using higher-fidelity 3D codes based on Computational Fluid Dynamics (CFD), instead.

Slug flow is a two-phase phenomenon of transient nature and therefore difficult to predict using analytical approaches [4]. Riser-induced slug flow is undesired in the process system because it leads to highly fluctuating flow and pressure levels, which can lead to poor separator performance among others [5]. The phenomenon of riser-induced slug flow is caused by the heavier liquid phase building up at geometrical low points in the riser, blocking the gas flow. Figure 1 shows the stages of a riser-induced slug cycle starting at top left. The cycle has a constant frequency which is influenced by the current flow parameters.

1. Slug formation—The slug starts forming at the bottom of the riser. As liquid blocks the pipe, the upstream gas pressure increases.

2. Slug production—As the slug reaches the top of the riser, liquid is transported to the separator.
3. Blowout—The compressed gas reaches the bottom of the riser as the gas pressure exceeds the gravitational head of the slug. The gas expands up the riser and causes an acceleration of the liquid slug.
4. Liquid fallback—Residual liquid falls to the bottom of the riser to restart the cycle.

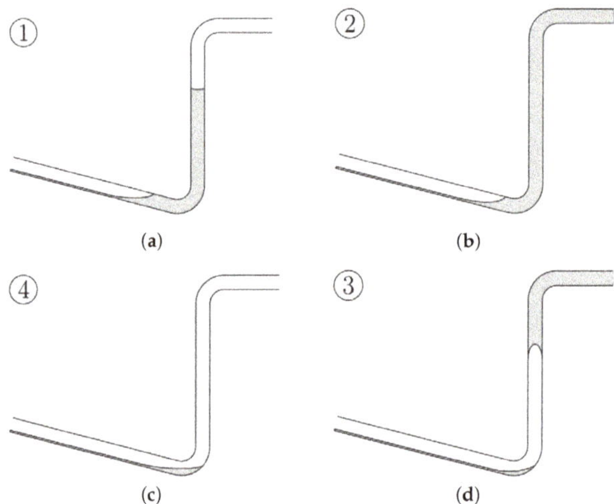

Figure 1. The stages of a riser-induced slug cycle. (**a**) Slug formation; (**b**) Slug production; (**c**) Liquid fallback; (**d**) Blowout.

The topic of gas-liquid slug flow have most recently been reviewed by Morgado et al. [6] who underlined the need for more experimental confirmation of numerical models. Presently, dynamic 1D models to predict the complex two phase flow pattern in offshore pipelines are based on the unit cell concept which was originally proposed by Wallis [7]. Research within these mechanistic models are focused on improving the closure relations to advance the prediction capabilities without significantly increasing the complexity of the solution [8]. This is most recently demonstrated by Soedarmo et al. [9] who suggested a new ad-hoc closure relationship for intermittent flow in pipes and by Fan et al. [10] who suggested a new model based on dimensional predictive regression. Models based on the unit cell concept are attractive as they only require very few computational resources, however it is well known that only CFD models can possibly correctly predict the complex physics of slug flow [11].

The numerical solution of the governing equations of fluid flow with spatial and temporal resolution of bubbles and turbulence is known as CFD. To model two-phase flow the Volume Of Fluid (VOF) method is most often used to resolve the gas-liquid interface. The VOF model have prove to be superior compared to other two-phase models for vertical flow [12] and have been validated for vertical slug flow using experimental data [13]. The VOF model have also been shown to be capable of predicting the full spectrum of flow pattern, such as stratified, wavy, slug, plug and annular flows [14,15]. Thus, the accurate simulation of the physics of the flow field using CFD can be used to give a better understanding of the transition between the different flow patterns [16,17]. 3D CFD simulations have also been used to directly provide input for the development of closure relationships [18]. Thus, the use of CFD enables the user to perform accurate predictions on the detailed dynamics of two-phase flow without the need to invest in expensive in-situ experimental tests [19]. 3D CFD simulations are able to predict non-symmetric nature of the interface shape and are thus able to capture flow phenomenon which cannot be modelled by 1D techniques [20]. CFD models for multiphase flow are also continuously developed to include new features such as Fluid Structure Interaction [21] and more accurate prediction of the gas liquid interface [22].

Previous studies have shown that there is a continuous efforts towards improving mechanistic models as well as CFD simulations using the VOF model. The aim of this study is to directly compare the prediction performance of these two models as well as their computational cost. By comparing objective parameters such as the predicted slug frequency and maximum pressure against newly acquired experimental data it is possible to make an informed choice whether to use high fidelity, but more computational intensive CFD models, compared to existing mechanistic 1D models.

In this study, two models have been developed and evaluated against experimental data acquired at the laboratory facility located in Aalborg University, Esbjerg campus. Models are developed through 1D approaches using the commercial software LedaFlow, while a more complex 3D CFD model is developed within the open source framework of OpenFOAM v6. The benefit of 1D modeling is the reduction in computational cost, while the appeal of 3D modeling is the potential improvement of modeling accuracy. Although the strong and weak points of both modeling principles are already known it is rarely quantified how big the deviations are in the key parameters characterizing the slug flow; this study focuses on comparing the models where the main evaluation is based on quantification of slug progression and period as well as the computational time.

In this article, the experimental facility and its setup is presented in Section 2. This includes physical properties of the facility and the given operating conditions of which the numerical models are based upon. In Section 3, the developed numerical models for both CFD and 1D simulations are addressed in terms of the chosen meshing strategies, discretization schemes and fluid and turbulence models. In Section 4, the numerical results and experimental data are presented and compared using pressure plots of the riser topside and low-point over time, together with the topside mass flow over time. The performance of the numerical models is evaluated based on the correlation with experimental data in terms of peak pressures and slug frequency, the ability to predict phase interfaces and their computational intensity. Finally, in Section 5, the article is concluded through summarizing the main findings of the present work.

2. Experimental Method

A process diagram for the test facility, including pipe section designations, is presented in Figure 2. Water is recirculated in the system by a pump while air is injected immediately after the recirculation pump and removed in the two-phase separator. The setup is more thoroughly described in Reference [23,24] while selected dimensions are presented in Table 1.

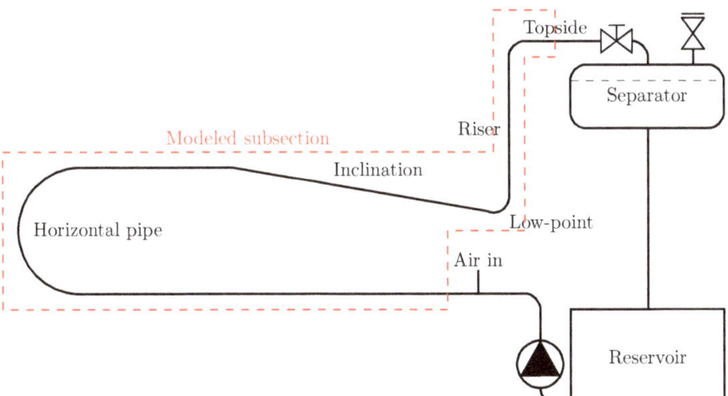

Figure 2. Overview of test setup.

Table 1. Dimensions of the test facility pipeline.

Parameter	Symbol	Value
Pipe material	-	PVC
Pipe diameter	D	0.051 m
Horizontal pipeline length	l_h	30 m
Inclination length	l_i	12 m
Riser length	l_r	6 m
Inclination angle	β	$-8°$
Material roughness	ε	1.5×10^{-6} m

Pressure and temperature transmitters are located along the pipes. The Siemens SITRANS P200 7MF1565 pressure transmitters at the low-point, p_{rb}, and topside, p_{rt}, are placed as presented in Figure 3. A bespoke Micro Motion ELITE Coriolis flow meter utilizing a CFM200M transmitter, capable of measuring both mass flow and density, is installed topside. The flow rate entering the separator is indicated with the symbol $\omega_{sep,in}$. The flow meter is developed for single-phase flow but is designed to reduce inaccuracy in two-phase flow. The topside choke valve is maintained fully open during the experiment. The actuators and transmitters are described in details in Reference [25].

Figure 3. Drawing of the well-pipeline-riser test facility.

The experimental configuration presented in Table 2 is based on Scenario 1, Test 1 of Reference [26]. The Table shows the input parameters and results of the laboratory test with averaged values as all parameters varies insignificantly with regards to change in fluid properties. The setpoint for the separator mass flow is 0.4 kg/s. The slug frequency is estimated from Fast Fourier Transform (FFT) analysis.

Table 2. Experimental test results. Volumes are stated for normal conditions.

Parameter	Symbol	Value
Separator pressure	p_{sep}	103 kPa
Temperature, avg.	T	22 °C
Liquid nominal avg. mass flow	$\omega_{L,in}$	0.4 kg/s
Gas nominal avg. volume flow	$\omega_{G,in}$	1.25 Nm^3/h
Topside valve opening	-	100%
Slug frequency	f_{slug}	1/78.5 Hz

The setup is replicated, and new data is recorded. This includes the low-point pressure and topside pressure and mass flow for the first 350 s after the initial transient slug cycle. The sampling rate is 100 Hz to ensure flow dynamics of interest is captured.

3. Numerical Methods

The software used for 1D modeling in this study is LedaFlow Engineering v2.4.255.030, referenced simply as LedaFlow. Alternative softwares with similar features include OLGA ; however, previous studies have shown that for the simulation of riser-induced slug flow, LedaFlow and OLGA perform comparably, with same strong and weak point [3]. In Figure 4, the LedaFlow simulation setup is presented using the LedaFlow interface.

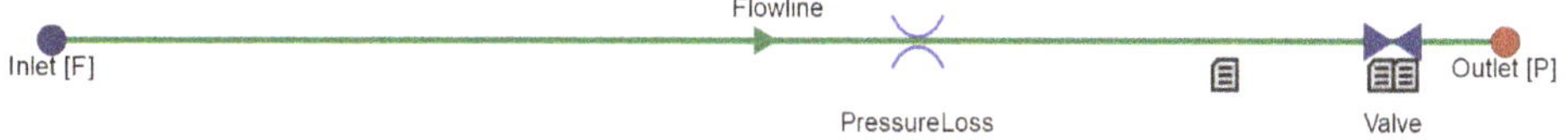

Figure 4. LedaFlow setup.

Each flow component requires specification of several parameters within the categories: INLET, OUTLET, PIPELINE, VALVE and LOSS. The parameters are listed in Table 3.

Table 3. Simulation model setup for LedaFlow.

	Parameter	Value	Unit
INLET	Mass flow rate, water	0.4	kg/s
	Mass flow rate, air	449×10^{-6}	kg/s
OUTLET	Pressure	103	kPa
	Temperature	295	K
	Volume fraction, water	0	-
	Volume fraction, air	1	-
PIPELINE	Thickness	7.75	mm
	Roughness	1.5×10^{-6}	m
	Heat capacity	1000	J/(kgK)
	Thermal conductivity	0.19	W/(mK)
	Density	1390	kg/m^3
VALVE	Opening	1	-
	Discharge coefficient	0.84	-
LOSS	k_L, 8° bend	0.6	-
	k_L, 98° bend w. $r/D = 10$	0.57	-
	k_L, 90° bend w. $r/D = 1.5$	0.3	-
	k_L, Globe valve, fully open	6.5	-

For Pressure Volume Temperature (PVT) settings, the user has the option to either use a PVT table or use constant PVT values defined by the user. As the experimental temperature change is insignificant and the total pressure change is not expected to exceed 2 bar, it is considered acceptable to use constant values instead of PVT tables. Thus, the constant values for air and water are implemented in LedaFlow.

3D Model

The solver used for this case is the standard solver `compressibleInterFoam` of OpenFOAM v6. The model is developed for two compressible, non-isothermal, immiscible fluids using the VOF method [27]. The model solves transport equations for continuity, the Reynolds Averaged Navier Stokes (RANS), volume fraction, turbulent kinetic energy, k, and the dissipation rate, ε, as they are implemented in the OpenFOAM v6 code.

For high mesh quality, a hexahedral butterfly mesh was developed, of which an excerpt is presented in Figure 5. A grid independency analysis was performed to determine an adequate mesh resolution evaluated based on interface resolution and the riser bottom pressure progress. The total cell count of the final mesh is 240 k.

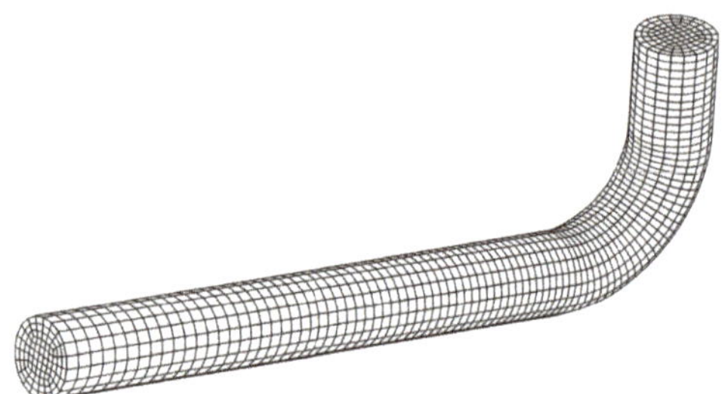

Figure 5. Cutout of model mesh near the low-point of the riser.

Four boundaries are specified; water inlet, air inlet, outlet and wall. Inlets are specified as velocity inlets with magnitudes equivalent to the mass flows specified in Section 2. A pressure outlet at atmospheric pressure is specified for the outlet boundary, while a no-slip condition is applied for the wall. The water void fractions are defined as constant 1 and 0 for the water and air inlets, respectively, while the outlet has an inlet/outlet condition, allowing reverse flow with a water void fraction of 0. Temperature conditions are constant values for inlets and wall, while a zero gradient is applied for the outlet. Finally, default settings for turbulence boundary conditions are applied.

The model utilizes the `compressibleInterFoam` solver with a 10^{-5} convergence criteria. Spatial discretization is obtained using 2nd order schemes, while the temporal discretization scheme is evenly weighted between 1st order implicit Euler and 2nd order Crank-Nicolson scheme to obtain convergence. The time step is controlled by limiting the Courant number for all cells to 0.8. The Realizable k-ε model with default model constants is used for turbulence modeling. From a sensitivity study, it was found that the model was insensitive to the choice of turbulence model when comparing formulations of the popular two-equation models, k-ε and k-ω. The Realizable k-ε model was chosen as this was the most stable of the models tested.

A grid independence study for the geometry has also been performed using 1st order upwind spatial discretization. The results for 1st order upwind are seen in Figure 6 and the results for 2nd order upwind are seen in Figure 7 for the same time step, represented as a contour plot of the water volume fraction during initial slug formation. In Figure 6, it can be seen that 1st order upwind gives a somewhat blurry prediction of the phase separation, which is not present for the simulation for 2nd order upwind in Figure 7. Because of this, it is evident that the simulation for 1st order upwind on the finest mesh is only similar to that for 2nd order upwind on the coarsest mesh. Thus, it is essential to use higher order discretization for slug flow simulations.

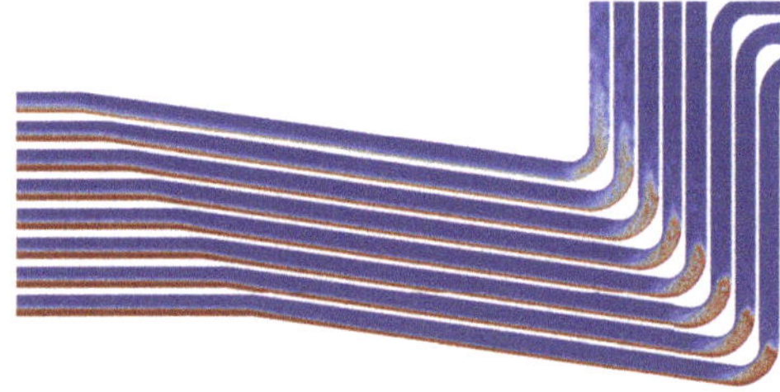

Figure 6. Longitudinal cut at $y = 0$ m. Coarsest mesh is at upper left and finest at lower right.

In Figure 7, when observing the interface boundaries of the solutions, smearing can be observed for the coarsest meshes due to their resolutions.

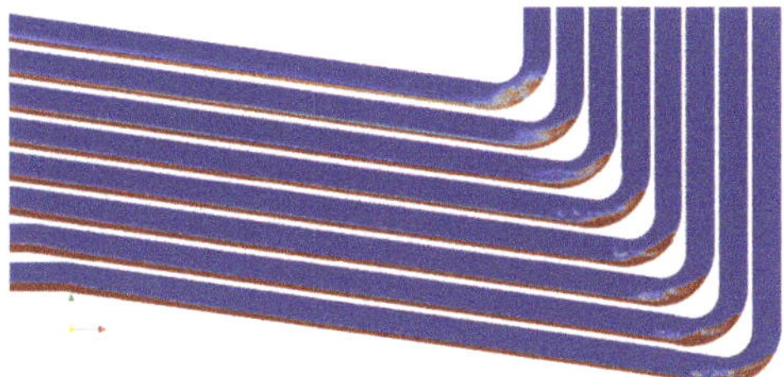

Figure 7. Longitudinal cut at $y = 0$ m. The coarsest mesh at upper left and finest at lower right.

4. Results and Discussion

Figure 8 shows the low-point pressure amplitudes. The results of the numerical models show good correlation in terms of both pressure levels and propagation, while the experimental data shows less pressure drop during a cycle.

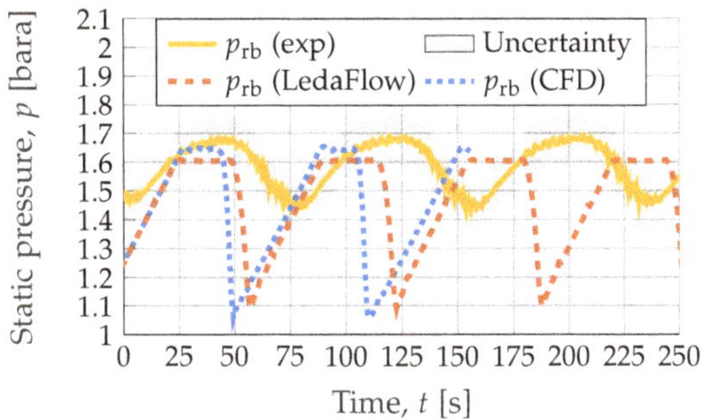

Figure 8. Comparison of Computational Fluid Dynamics (CFD), LedaFlow and experimental results for the low-point pressures.

The difference in low-point pressure ranges is explainable by the extent the riser is emptied during blowout. Both numerical models predict that the riser is emptied during blowout. This explains the drop to atmospheric pressure as there is essentially no water column to add hydrostatic pressure. When physically observing the experiment execution, the air is evacuated from the riser through smaller Taylor bubbles flowing through a water column, adding a hydrostatic pressure component.

Figure 9 compares the topside pressures. By observing the plot, it is evident that the experimental pressure data is above atmospheric during the entire sampling period, which none of the numerical simulation models could reproduce.

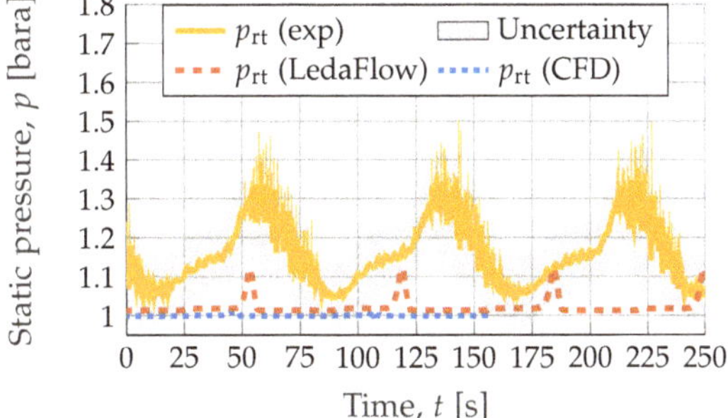

Figure 9. Comparison of CFD, LedaFlow and experimental results for the topside pressures.

Numerical models predict atmospheric pressure levels besides during the blowout phase, where a small pressure peak is observed, for example, at 50 s. The peak in the CFD results are barely observable from the plot, while the peak of 1D transient model is clearly distinguishable. The higher peak pressure of the LedaFlow model, compared to the CFD model, is caused by the implementation of the topside valve in the LedaFlow model, which has not been accounted for in the CFD model.

Figure 10 compares the topside mass flows. The experimental mass flow fluctuates in the range of 0–0.5 kg/s. Observing the numerical results, the mass flows are within this range, except for momentary peaks at levels of around 3 and 5 kg/s for LedaFlow and CFD, respectively. These peaks represent the blowout phase, followed by a slug formation phase where the mass flow rates reduce to approximately 0 kg/s. This is due to pure air flow at the topside position during slug formation. During liquid production, the mass flow increases to around 0.5 kg/s. These cyclic phases are clearly separated in the numerical data.

It is harder to indicate the cycle phase occurrences in the experimental data. At the liquid production phase, observable at around 75 s, numerical and experimental data shows good correlation. The blowout, fallback and slug formation phases are harder to distinguish. A small peak can be observed at 210 s, indicating a blowout, after which the mass flow decreases slowly while fluctuating, which might indicate two-phase flow topside during the remaining cycle phases. This is supported by physical observations of the experiment execution, where only the top of the riser is emptied of water, while air is evacuated from the riser afterwards through Taylor bubbles.

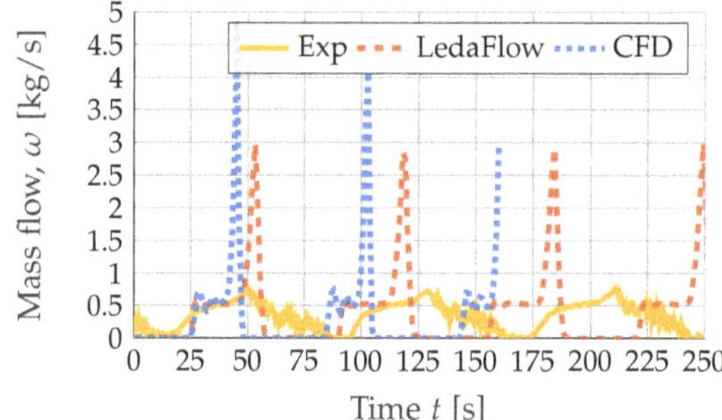

Figure 10. Comparison of CFD, LedaFlow and experimental results for the topside mass flow rate.

The high mass flow peaks of the numerical models are believed to be evidence of a more severe blowout predicted by the models compared to experimental data. This also explains the slower decrease of mass flow in the experimental data.

The inability to predict the experimentally observed blowout using the numerical models is puzzling. It is believed that the specification of the outlet pressure is the main cause of this disagreement. The outlet is specified to be at atmospheric pressure, which is justifiable, as the downstream separator is vented to the atmosphere. From observing the resulting pressure amplitudes of the model, this boundary condition does not seem to capture the damping possessed by the physical system in the current models. Also, the outlet boundary conditions might be cause of the difference in topside pressure when comparing model data to experimental.

Figures 8–10 indicate a difference in slug periods which is quantified through FFT analysis of the low-point pressure data. Normalized FFT plots are presented in Figure 11 and a significant difference between the data sets is evident. The model with the best prediction of the actual slug frequency is LedaFlow, which predicts a period of 66 s compared to 78.5 s of experimental data, while the CFD model results predicts a period of 61 s. This difference is believed to be attributable to the difference in blowout progression.

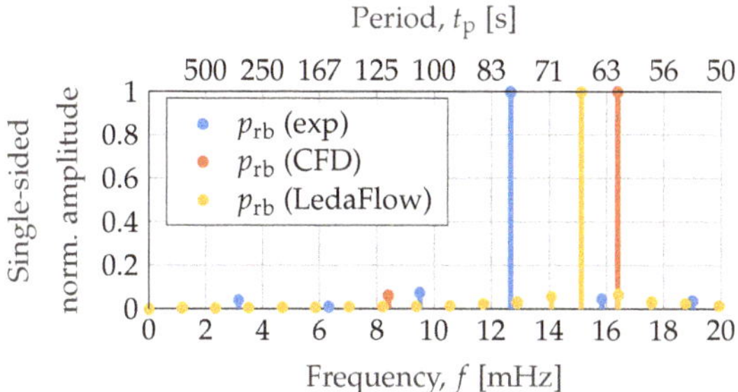

Figure 11. Comparison of Fast Fourier Transform (FFT) spectra of low-point pressures.

The models are compared and evaluated in Table 4.

Table 4. Comparison of 1D and 3D models.

Parameter	3D CFD	1D Transient
Cell count	240,550	165
Run time	57 h 30 min	11 min
Processor use	7 cores, 3.5 GHz	Serial, 2.5 GHz
Simulated time	330 s	1000 s
Predictions:		
- Pressure range	Comparable	
- Mass flow rate	Comparable	
- Slug frequency	22.3% deviation	15.7% deviation
- Interface prediction	Good	Poor

The LedaFlow and CFD models perform similarly in predicting the pressure ranges and mass flow rate. For the slug period, the LedaFlow model performs slightly better than the CFD model when comparing to experimental data. The advantage of the CFD model is found to be the ability to predict phase interfaces as shown in Figure 12 for a Taylor bubble.

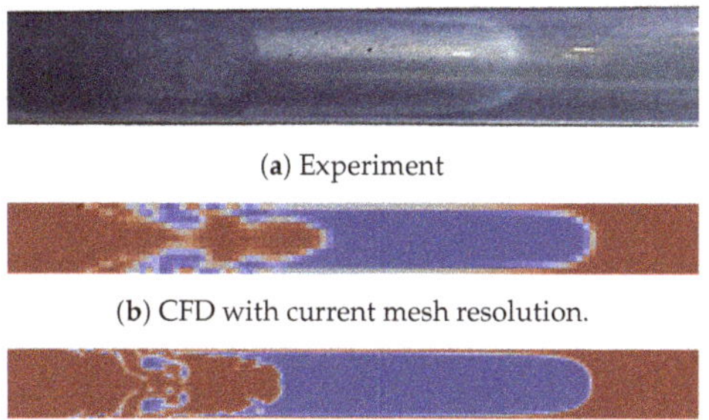

(a) Experiment

(b) CFD with current mesh resolution.

(c) CFD with increased mesh resolution (cell center spacing of 2 mm).

Figure 12. Comparison of Taylor bubble observed in the riser in experiment and CFD. The pictures are rotated 90° clockwise.

The runtime of LedaFlow is 1‰ that of the CFD model to simulate 330 s, where LedaFlow is run on a single core 2.5 GHz laptop and the CFD model is decomposed to utilize 7 cores on a 3.5 GHz workstation. Thus, the choice of simulation model strongly depends on the desired accuracy of the interfaces and the available time, as current models perform comparable for cycle prediction in terms of pressure levels and cycle frequency.

5. Conclusions

During this study, numerical models have been developed for predicting the flow features of a water-air flow through an experimental riser pipeline setup under operating conditions producing riser-induced slug flow. A 3D CFD model was developed based on the VOF model using RANS with the realizable k-ε turbulence model. The model was developed within the OpenFOAM v6 framework. In addition, a 1D transient model was developed in the commercial software LedaFlow to compare options for numerical simulation of the case.

Based on the case study's results from the laboratory testing facility, where the models' accuracy and computational time have been quantified and presented, the following can be concluded:

- Both commercial 1D and 3D CFD modeling can be used to simulate riser-induced slugs if the main focus is slug progression and period.
- Both developed models predicted maximum pressure levels close to experimental.
- The developed models predicted slug periods of 61 s and 66 s, while the experimental was found to be 78 s.
- The performance of the 1D and 3D models are comparable in terms of predicting pressure levels and slug frequencies.
- The 1D model is far less computationally intensive compared to 3D CFD.
- 3D CFD modeling proves superior to 1D codes in resolving phase interfaces.
- Model accuracy is expected to be further improvable by revising boundary conditions to better capture the natural damping of the physical system.
- If the selected key evaluation criteria is slug amplitude and frequency (such as for most anti-slug control schemes) both LedaFlow and 3D CFD can be used, while a focus on phase interfaces demands 3D CFD modeling.

Author Contributions: Author Contributions: Manuscript preparation, R.T.J. and G.R.T.; Reviewing and editing, R.T.J., G.R.T., M.M. and S.P.; Supervision, M.M. and S.P. All authors have read and agreed to the published version of the manuscript.

Funding: This research received no external funding.

Conflicts of Interest: The authors declare no conflict of interest.

References

1. Pedersen, S.; Durdevic, P.; Yang, Z. Review of Slug Detection, Modeling and Control Techniques for Offshore Oil & Gas Production Processes. *IFAC-PapersOnLine* **2015**, *48*, 89–96.
2. Pedersen, S.; Durdevic, P.; Yang, Z. Challenges in slug modeling and control for offshore oil and gas productions: A review study. *Int. J. Multiph. Flow* **2017**, *88*, 270–284. [CrossRef]
3. Belt, R.; Duret, E.; Larrey, D.; Djoric, B.; Kalali, S. Comparison of Commercial Multiphase Flow Simulators with Experimental and Field Databases. In Proceedings of the *15th International Conference on Multiphase Production Technology, Cannes, France, 15–17 June 2011*; BHR Group: Bedford, UK, 2011.
4. Taitel, Y.; Barnea, D. Two-Phase Slug Flow. In *Advances in Heat Transfer*; Hartnett, J.P., Irvine, T.F., Eds.; Elsevier: Amsterdam, The Netherlands, 1990; Volume 20, pp. 83–132.
5. Hill, T.; Wood, D. Slug Flow: Occurrence, Consequences, and Prediction. In Proceedings of the University of Tulsa Centennial Petroleum Engineering Symposium, Tulsa, OK, USA, 29–31 August 1994.
6. Morgado, A.; Miranda, J.; Araujo, J.; Campos, J. Review on vertical gas–liquid slug flow. *Int. J. Multiph. Flow* **2016**, *85*, 348–368. [CrossRef]
7. Wallis, G.B. *One-Dimensional Two-Phase Flow*; McGraw-Hill: New York, NY, USA, 1969.
8. Adames, P.; Smith, S.; Young, B.; Svrcek, W. Recent advances in slug flow modelling. In *6th North American Conference on Multiphase Technology, Banff, AB, Canada, 4–6 June 2008*; BHR Group: Bedford, UK, 2008.
9. Soedarmo, A.; Fan, Y.; Pereyra, E.; Sarica, C. A unit cell model for gas-liquid pseudo-slug flow in pipes. *J. Nat. Gas Sci. Eng.* **2018**, *60*, 125–143. [CrossRef]

10. Fan, Y.; Al-Safran, E.; Pereyra, E.; Sarica, C. Modeling pseudo-slugs liquid holdup in slightly upward inclined pipes. *J. Pet. Sci. Eng.* **2020**, in press. [CrossRef]
11. Montoya, G.; Lucas, D.; Baglietto, E.; Liao, Y. A review on mechanisms and models for the churn-turbulent flow regime. *Chem. Eng. Sci.* **2016**, *141*, 86–103. [CrossRef]
12. Guerrero, E.; Munoz, F.; Ratkovich, N. Comparison between Eulerian and VOF models for two-phase flow assessment in vertical pipes. *Cienc. Tecnol. Y Futuro* **2017**, *7*, 73–84. [CrossRef]
13. Tomiyama, A.; Zun, I.; Sou, A.; Sakaguchi, T. Numerical analysis of bubble motion with the VOF method. *Nucl. Eng. Des.* **1993**, *141*, 69–82. [CrossRef]
14. Vaze, M.J.; Banerjee, J. Experimental visualization of two-phase flow patterns and transition from stratified to slug flow. *Proc. Inst. Mech. Eng. Part C J. Mech. Eng. Sci.* **2011**, *225*, 382–389. [CrossRef]
15. Thaker, J.P.; Banerjee, J. CFD Simulation of Two-Phase Flow Phenomena in Horizontal Pipelines Using OpenFOAM. In Proceedings of the Fortieth National Conference on Fluid Mechanics and Fluid Power, Hamirpur, India, 12–14 December 2013.
16. Hernandez-Perez, V.; Abdulkadir, M.; Azzopardi, B.J. Grid Generation Issues in the CFD Modelling of Two-Phase Flow in a Pipe. *J. Comput. Multiph. Flows* **2011**, *3*, 13–26. [CrossRef]
17. Li, Y.; Hosseini, M.; Arasteh, H.; Toghraie, D.; Rostami, S. Transition simulation of two-phase intermittent slug flow characteristics in oil and gas pipelines. *Int. Commun. Heat Mass Transf.* **2020**, *113*. [CrossRef]
18. Valdes, J.; Pico, P.; Pereyra, E.; Ratkovich, N. Evaluation of drift-velocity closure relationships for highly viscous liquid-air slug flow in horizontal pipes through 3D CFD modelling. *Chem. Eng. Sci.* **2020**, *217*. [CrossRef]
19. Pineda-Perez, H.; Kim, T.; Pereyra, E.; Ratkovich, N. CFD modeling of air and highly viscous liquid two-phase slug flow in horizontal pipes. *Chem. Eng. Res. Des.* **2018**, *136*, 638–653. [CrossRef]
20. Tang, K.; Ouyang, Y.; Agarwal, R.; Chen, J.; Xiang, Y.; Chen, J. Computation of gas-liquid flow in a square bubble column with Wray-Agarwal one-equation turbulence model. *Chem. Eng. Sci.* **2020**, *218*. [CrossRef]
21. Mohmmed, A.; Al-Kayiem, H.; Osman, A.; Sabir, O. One-way coupled fluid–structure interaction of gas–liquid slug flow in a horizontal pipe: Experiments and simulations. *J. Fluids Struct.* **2020**, *97*. [CrossRef]
22. Li, Q.; Dong, F.; Li, L. An interface-sharpening method with adaptive mesh refinement for volume-of-fluid simulations of two-phase compressible flows. *Comput. Fluids* **2020**, in press. [CrossRef]
23. Pedersen, S. Plant-Wide Anti-Slug Control for Offshore Oil and Gas Processes. Ph.D. Thesis, Aalborg University, Aalborg, Denmark, 2016.
24. Biltoft, J.; Hansen, L.; Pedersen, S.; Yang, Z. Recreating Riser Slugging Flow Based on an Economic Lab-sized Setup. *IFAC Proc. Vol.* **2013**, *46*, 47–52. [CrossRef]
25. Pedersen, S.; Jahanshahi, E.; Yang, Z.; Skogestad, S. Comparison of Model-Based Control Solutions for Severe Riser-Induced Slugs. *Energies* **2017**, *10*, 2014. [CrossRef]
26. Pedersen, S.; Durdevic, P.; Yang, Z. Influence of riser-induced slugs on the downstream separation processes. *J. Pet. Sci. Eng.* **2017**, *154*, 337–343. [CrossRef]
27. The OpenFOAM Foundation. *OpenFOAM User Guide version 6*; The OpenFOAM Foundation: London, UK, 2018.

Article

Modification of Interaction Forces between Smoke and Evacuees

Sungryong Bae [1], Jun-Ho Choi [2] and Hong Sun Ryou [3],*

[1] Department of Advanced Industry Convergence, Chosun University, Gwangju 61452, Korea; sbae@chosun.ac.kr

[2] Division of Architectural and Fire Protection Engineering, Pukyong National University, Busan 48513, Korea; jchoi@pknu.ac.kr

[3] School of Mechanical Engineering, Chung-Ang University, Seoul 06974, Korea

* Correspondence: cfdmec@cau.ac.kr; Tel.: +82-2-813-3669

Received: 7 July 2020; Accepted: 11 August 2020; Published: 13 August 2020

Abstract: The most used fire effect models on evacuees are only focused on the physical capacity of the evacuees. However, some of the evacuees in a fire situation continuously move through the familiar route, although the familiar route is smoke-filled and they know that they are moving towards the fire source. Thus, the additional evacuation models are required for considering the behavioral changes due to the psychological pressure when the evacuees are moving through the smoke or towards the fire source. In this study, the inner smoke region force is modified to improve the accuracy and practicality of the BR-smoke model by varying the walking speed according to the smoke density. Additionally, the BR-smoke model is applied to FDS+Evac to compare the simulation results of the modified BR-smoke model with those of existing models. Based on the results, the evacuation characteristics inside the smoke region can be improved by using the modified BR-smoke model because the evacuees are continuously influenced by the modified inner smoke force inside the smoke region. However, additional studies for determining more reliable evacuee psychological factors are required to improve the reality of the modified BR-smoke model.

Keywords: evacuation; interaction between smoke and evacuees; inner smoke force; modified BR-smoke model

---

1. Introduction

Recently, the performance-based design of buildings has been regulated to reduce property damage and casualties in building fire situations [1]. Additionally, assessing building evacuation characteristics is highly important for decreasing the number of casualties and improving evacuation safety in fire situations. Most evacuation models apply various walking speeds according to the smoke density [2,3] and a suffocation effect according to the amount of toxic gases from the fire [4] when considering the interaction between a fire and the evacuees. These models focus on the variation in the physical capacities of the evacuees such as their walking speed and whether they live or die. Thus, these models cannot estimate behavioral changes due to the psychological pressure experienced when the evacuees are confronted with a fire.

Moreover, it is assumed that the evacuees do not move forward to the fire and smoke region in a fire situation. Thus, those two models for walking speed [2,3] and toxic effect [4] are suitable to the fire simulation with this assumption. However, the evacuees in a fire situation would get into a panic and they should move based on their instincts because they cannot make a rational decision [5]. Especially, the evacuees in a panic continuously move through the familiar route according to the homing instinct, although the familiar route is smoke-filled and they know that they are moving towards the fire source [6–8]. Therefore, the evacuees who are finishing the evacuation by moving through the smoke or around the fire source have to be considered in the evacuation scenario of the fire simulation. However, as previously described, the models for the fire effect on the evacuees are only focused on the physical capacity of the evacuees, and these models cannot consider the behavioral changes due to the psychological pressure when the evacuees are moving through the smoke or towards the fire source.

Recently, Bae and Ryou [9] suggested a force-based model (BR-radiation model) for considering behavior changes due to the pressure experienced when evacuees feel the radiative heat flux from a fire. They converted the psychological pressure from the radiative heat flux into the radiation force, which causes the evacuees to move while avoiding the fire source. Then, they compared the new evacuation model with existing models and identified that the reliability of the evacuation characteristics can be improved by using their proposed evacuation model. Afterwards, Bae et al. [10] performed an experimental study to improve the practicality of the simulated interaction between evacuees and a fire via radiation. The reliability and practicality of the BR-radiation model was further improved by applying experimental results to the BR-radiation model.

Bae and Ryou [11] suggested another force-based model (BR-smoke model) for considering behavioral changes due to the pressure experienced when evacuees are confronted by the smoke region. The smoke force, which represents the psychological pressure from the smoke region, is divided into an outer and inner smoke region force to improve model accuracy and practicality. They also identified an improvement in the reliability of the evacuation characteristics by comparing the proposed model with existing models. However, the reliability of the BR-smoke model was diminished by the inner smoke region force because it was assumed that the inner smoke region force was directly proportional to the visibility at the position of the evacuees. Hence, the inner smoke region force decreased as the smoke density increased, although the probability of a direction change of the evacuees increases as the smoke density is increased [12]. Thus, a modification of the inner smoke region force is required to improve the reliability of the BR-smoke model.

Therefore, in this study, the inner smoke region force of the BR-smoke model is modified by the varying walking speed according to the smoke density [3]. In addition, the BR-smoke model is applied to FDS+Evac [13], which can simultaneously calculate the fire and evacuation simulations. Then, the simulation results of the modified BR-smoke model are compared with those of existing models to verify the reliability of the BR-smoke model.

2. Modification of the BR-Smoke Model

2.1. Helbing's Movement Model

The Helbing's movement model [14] is one of the widely used models for estimating the movement of the evacuee, and it can be categorized as the force-based model. That is, the Helbing's movement model was developed by substituting the psychological pressures experienced by the surrounding people or environment into the interaction forces between evacuee and evacuees or walls, respectively. The movement characteristics are then estimated by using the mathematically modeled forces (social force, contact force, and attraction force). The Helbing's movement model is represented as follows:

$$m_i \frac{d^2 \mathbf{x}_i}{dt^2} = \frac{m_i}{\tau_i}\left(\mathbf{v}_i^o + \mathbf{v}_i\right) + \sum_{i \neq j}\left(\mathbf{f}_{ij}^{soc} + \mathbf{f}_{ij}^{cont} + \mathbf{v}_{ij}^{att}\right) + \sum_{W}\left(\mathbf{f}_{iw}^{soc} + \mathbf{f}_{iw}^{cont}\right) \tag{1}$$

where $\mathbf{x}_i(t)$ is the position vector of the agent, $\mathbf{v}_i^o(t)$ is the desired walking speed of the agent, $\mathbf{v}_i(t)$ is the estimated walking speed of the agent, τ_i is a certain characteristic time for each evacuee, and $\mathbf{f}^{soc}$, $\mathbf{f}^{cont}$, and $\mathbf{f}^{att}$ are the social force, contact force, and attraction force, respectively.

However, as represented in Equation (1), the Helbing's movement model cannot consider the fire effect on the evacuation movement, because the fire effects which come from the fire source and the smoke are not included in the model. Therefore, an additional model for considering the fire effect on the evacuees should be required for improving the accuracy of the evacuation simulation results in a fire situation.

2.2. Existing Smoke Force Model

As previously described, Bae and Ryou [11] suggested the force-based model, BR-smoke model, for considering the effect of the smoke from the fire on the evacuees. In the BR-smoke model, the forces for an outer/inner smoke region were included for considering the psychological pressures experienced when the evacuees are confronted with the smoke or they move through the smoke, respectively. The BR-smoke force is represented as follows:

$$\mathbf{f}_{iS} = \mathbf{f}_{iS}^{outer} + \mathbf{f}_{iS}^{inner} = S_{iS}\exp\left(-d_i^S\right)u\left(d_i^S\right)\mathbf{n}_{iS} + S_{iS}\frac{V_i}{V_\infty}\left(1 - u\left(d_i^S\right)\right)\mathbf{n}_{iS} \tag{2}$$

where $\mathbf{f}_{iS}$, $\mathbf{f}_{iS}^{outer}$, and $\mathbf{f}_{iS}^{inner}$ are the BR-smoke force, the outer smoke region force, and the inner smoke region force, respectively. Moreover, d_i^S is the minimum distance between the human and the smoke, $\mathbf{n}_{iS}$ is the unit vector pointing from the smoke boundary to the human, S_{iS} is the maximum value, 125 N, of the BR-smoke force, $u(d_i^S)$ is the unit step function at $d_i^S = 0$, and V_i and V_∞ are the visibility at the position of the evacuees and the maximum visibility, 30 m, respectively.

The BR-smoke model was then suggested by applying the BR-smoke force on the Helbing's movement model. After that, they verified that the reliability of the evacuation simulation in the smoke-filled area can be improved by applying the BR-smoke model.

2.3. Modification of the BR-Smoke Model

As previously described, the inner smoke region force represents the pressures experienced when evacuees move through the smoke region. This force acts on the evacuees to induce them to seek egress from the smoke region for their safety. However, as represented in Equation (2), it is assumed that the inner smoke region force is proportional to the visibility at the position of the evacuees. Thus, the magnitude of the inner smoke region force decreases as the evacuees move towards the dense smoke region.

Therefore, the inner smoke region force needs to be modified to improve the reliability of the BR-smoke force. In this study, it is assumed that the inner smoke region force is directly proportional to the variation in smoke density according to the time. In addition, the model representing the variation in the walking speed according to the smoke density [3] is used to determine the inner smoke region force that satisfies the above assumption. That is, the acceleration of evacuees in a smoke region is derived by differentiating the Frantzich and Nilsson correlation by time. Then, the acceleration is represented as follows:

$$\frac{dv_i^o}{dt} = \frac{d}{dt}\left(v_i^o + K_s v_i^o \frac{\beta}{\alpha}\right) = v_i^o \frac{\beta}{\alpha}\frac{dK_s}{dt} \tag{3}$$

where v_i^o is the desired speed of the evacuees, K_s is the extinction coefficient ($[K_s] = \text{m}^{-1}$), and α and β are the coefficients 0.706 ms^{-1} and -0.057 m^2 s^{-1}, respectively [3].

Then, the magnitude of the inner smoke region force can be derived by multiplying the evacuee's mass by the acceleration, and the magnitude of the inner smoke region is represented as follows:

$$\mathbf{f}_{iS}^{inner} = m_i a_i = m_i \frac{dv_i^o}{dt}\mathbf{n}_{iS} = m_i v_i^o \frac{\beta}{\alpha}\frac{dK_s}{dt}\mathbf{n}_{iS} \tag{4}$$

where m_i is the mass of the evacuees. Moreover, the direction of the inner smoke region force, $\mathbf{n}_{iS}$, is assumed to be the inner product of the opposite walking direction and the minimum distance direction between the evacuee and boundary of the smoke region.

Furthermore, the individual psychological pressure outside of the smoke region cannot be considered with the existing BR-smoke model, because the maximum value of the BR-smoke force was specified as a constant value, 125 N, as described in Section 2.2. Thus, the maximum value of the outer smoke region force, S_{iS}, is also modified by applying the maximum magnitude of the individual motive force, $m_i v_i^o / \tau_i$.

Then, based on these modifications, the modified smoke force of the BR-smoke model is represented as follows:

$$\mathbf{f}_{iS} = \mathbf{f}_{iS}^{outer} + \mathbf{f}_{iS}^{inner} = \frac{m_i v_i^o}{\tau_i}\exp(-d_{is})\mathrm{u}(d_{is})\mathbf{n}_{iS} + m_i v_i^o \frac{\beta}{\alpha}\frac{dK_s}{dt}(1 - \mathrm{u}(d_{is}))\mathbf{n}_{iS} \tag{5}$$

Finally, the modified smoke force is applied to the force-based evacuation model, Helbing's social force model [14], and the modified BR-smoke model is as follows:

$$m_i \frac{d^2\mathbf{x}_i}{dt^2} = \frac{m_i}{\tau_i}\left(\mathbf{v}_i^o + \mathbf{v}_i\right) + \sum_{i \neq j}\left(\mathbf{f}_{ij}^{soc} + \mathbf{f}_{ij}^{cont} + \mathbf{v}_{ij}^{att}\right) + \sum_{W}\left(\mathbf{f}_{iw}^{soc} + \mathbf{f}_{iw}^{cont}\right) + \mathbf{f}_{iS} \tag{6}$$

The modified BR-smoke model is applied to FDS+Evac [14], which can simultaneously calculate the fire and evacuation simulations. Then, the evacuation characteristics and forces acting on the evacuees are compared with the correlation of Frantzich and Nilsson and the BR-smoke model.

3. Simulation Conditions

Figure 1 represents the computational domain and desired direction for every grid point. As shown in Figure 1a, the arbitrary space that has dimensions of 24 m (L) × 16 m (W) × 2.4 m (H) is used to analyze the evacuation characteristics. The fire source is positioned at the center of the computational domain, and the size of the fire is assumed to be 2.0 MW. Additionally, the fire growth is ignored to maximize smoke generation and spreading.

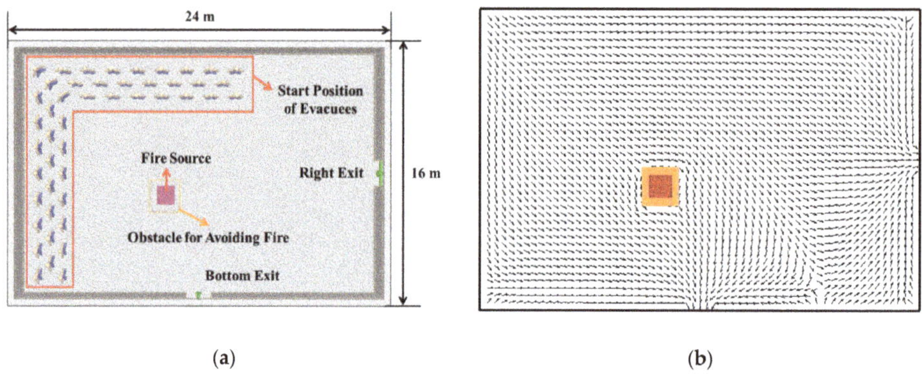

(a)

(b)

Figure 1. The (a) computational domain and (b) desired direction of every grid point.

Forty-seven adult males are positioned along the left and top walls of the computational domain to cause the evacuees to be influenced by the smoke region during evacuation. As represented in Figure 1a, the initial position and direction of the evacuees are all controlled for minimizing the random effect on the simulation results for each case. Moreover, all the evacuees immediately begin to evacuate after the fire occurs to the exits located at the center of the bottom and right walls. In FDS+Evac, the desired direction of the evacuees is determined by the potential flow field of the computational domain [13]. If the path for bypassing the fire is not specified, some evacuees move through the fire source. Therefore, an arbitrary obstruction is placed around the fire for evacuees to bypass the fire in the evacuation domain. The desired direction for all positions in the computational domain is shown in Figure 1b.

Table 1 lists the simulation cases according to the evacuation model. As listed in Table 1, all conditions, except for the movement model, are the same for each case. In Case 1, the Helbing's social force model [14] is used to analyze the evacuation characteristics, and the Frantzich and Nilsson correlation [3] is applied to consider the interaction between the fire and evacuees. In Case 2, the BR-model suggested by Bae and Ryou [11] is used to analyze the evacuation characteristics. In Case 3, the modified BR-model is used to analyze the evacuation characteristics. Additionally, in Cases 2 and 3, the Frantzich and Nilsson correlation [3] is applied to consider the walking variation according to the smoke density. Moreover, the random force acting on the evacuees are ignored for minimizing the random effect on the simulation results for each case. Then, the evacuation characteristics are analyzed by comparing forces (motive force, social force, and smoke force) and walking speed.

Table 1. The cases for the fire and evacuation simulations.

	Case 1	Case 2	Case 3
Movement Model	Social force model [14]	Existing BR-smoke model [11]	Modified BR-smoke model
Fire Grid Size	0.2 m		
Fire Growth	Ignored		
Fire Size	2 MW (toluene)		
No. of Evacuees	47 males		
Evacuation Grid Size	0.4 m		
Variation of Walking Speed	Correlation from Frantzich and Nilsson [3]		

4. Results and Discussion

Figures 2 and 3 represent the distribution of the evacuees at 2 and 4 s, respectively. As represented in Figure 2, the distributions of the evacuees at 2 s are nearly the same between the cases because the smoke region is farther away from the evacuees and can influence them less. However, as represented in Figure 3, the distributions of the evacuees after 4 s are significantly different from each other because the evacuees are confronted by the smoke region and move through it after 4 s. Some evacuees for Case 1 at 4 s walk into the smoke region because the interaction between the evacuees and the smoke region is not considered in the evacuation simulation. In contrast, the evacuees for Cases 1 and 2 at 4 s hesitate to walk into the smoke region, and they are distributed around the boundary of the smoke region because of the influence of the outer smoke region force.

Figure 4 represents the forces acting on a representative evacuee and the walking speed when the representative evacuee is confronted by the smoke region. An evacuee initially positioned nearby the upper left corner is selected as the representative evacuee because it is assumed that this evacuee is evacuated last from the computation domain for every case. The motive force applied in FDS+Evac for simulating the evacuation is used for making the evacuees move along the desired direction as predetermined by the potential flow. Moreover, the social force is used for making the evacuees change the walking speed and direction when they are confronted with the other evacuees or obstacles.

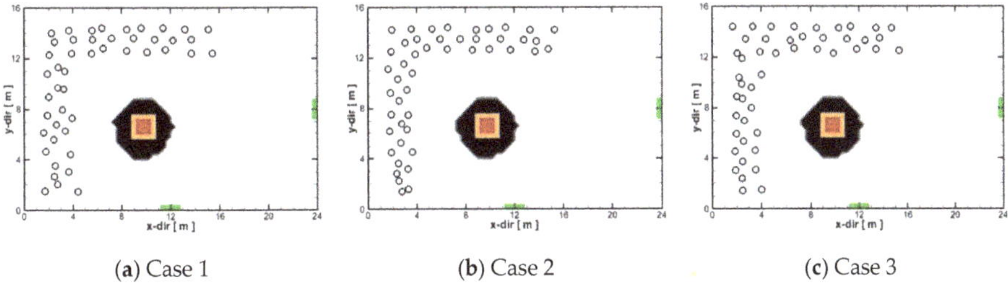

(a) Case 1   (b) Case 2   (c) Case 3

Figure 2. The distribution of the evacuees at 2 s.

(**a**) Case 1 (**b**) Case 2 (**c**) Case 3

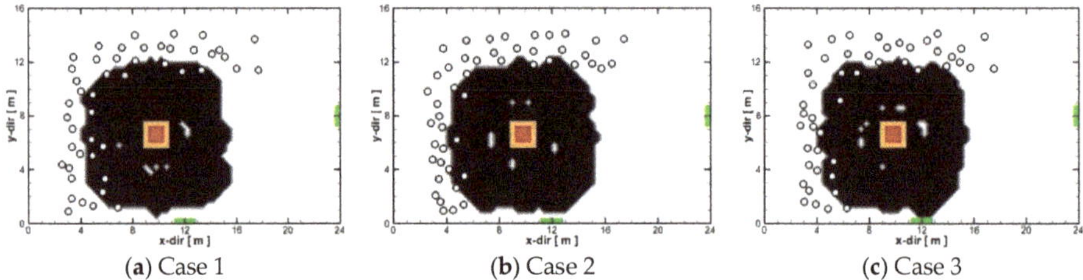

Figure 3. The distribution of the evacuees at 4 s.

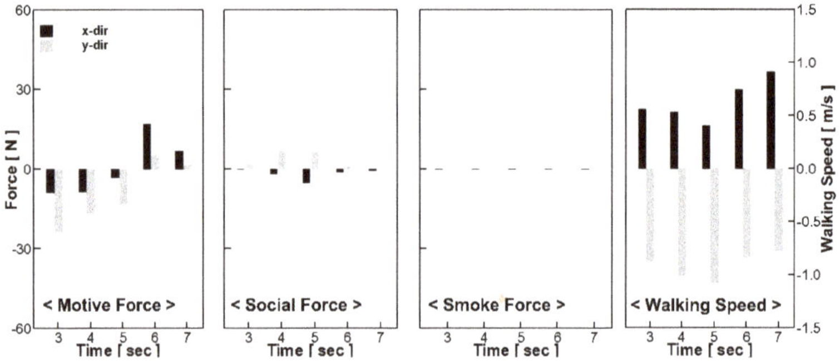

(**a**) Case 1

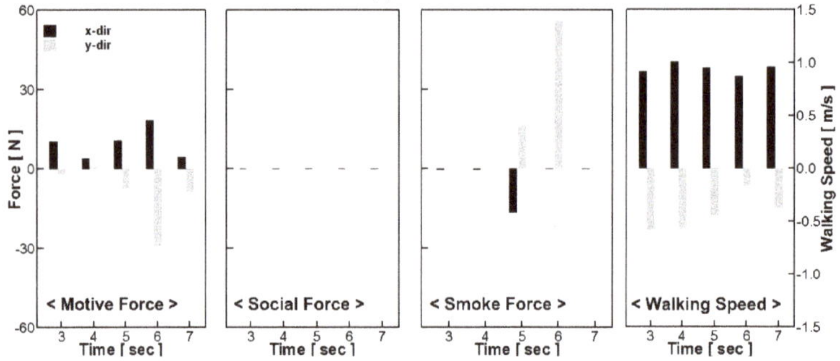

(**b**) Case 2

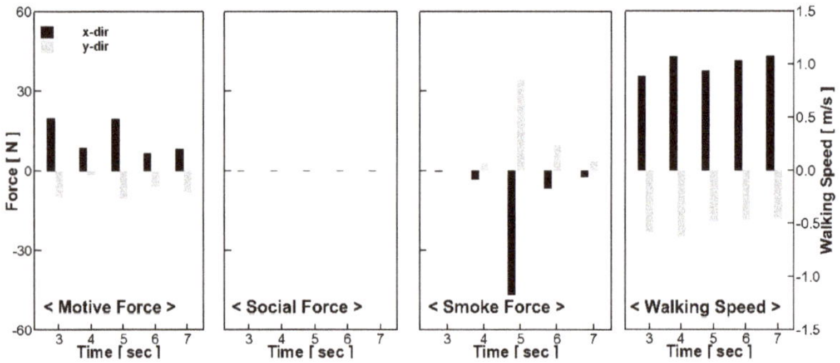

(**c**) Case 3

Figure 4. The forces acting on the representative evacuee and the walking speed from 3 to 7 s.

As represented in Figure 4a, the representative evacuee for Case 1 changes its walking speed only according to the influence of the motive and social forces. Based on these acting forces, the representative evacuee for Case 1 initially moves along the predetermined direction and changes its walking speed and direction after being confronted with the other evacuees. That is, it is confirmed that the behavioral changes of the evacuees due to the smoke cannot be considered with the Helbing's movement model.

In contrast, the representative evacuees for Cases 2 and 3 change their walking directions and decrease their walking speed because they are influenced by the outer smoke region force when confronted by the smoke region (Figure 4b,c). In addition, the social forces for Cases 2 and 3 do not act on the representative evacuee, in contrast with Case 1, because the representative evacuees for Cases 2 and 3 are further away from the other evacuees due to the influence of the outer smoke region force at around 5 s. That is, it is confirmed that the behavioral changes of the evacuees due to the smoke can be well considered by applying the BR-smoke model.

Figures 5 and 6 represent the distributions of the evacuees at 8 and 12 s, respectively. As represented in Figure 5, the evacuees in Case 1 that are moving towards the right exit are further away from the fire in contrast with those moving towards the bottom exit. In addition, the distributions for Cases 2 and 3 are extremely similar to each other at this time because the distribution is estimated by the influence of the outer smoke region force. Moreover, as represented in Figure 6, the distributions at 12 s are somewhat different from each other. The evacuees in Case 2 are congregated around the center of the computational domain in the y-direction because the evacuees only move towards the exit with the motive force influence. In contrast, the evacuees in Case 3 are further away from the fire while maintaining the distribution of that at 8 s because the evacuees are continuously changing their walking direction because of the influence of the modified inner smoke region force.

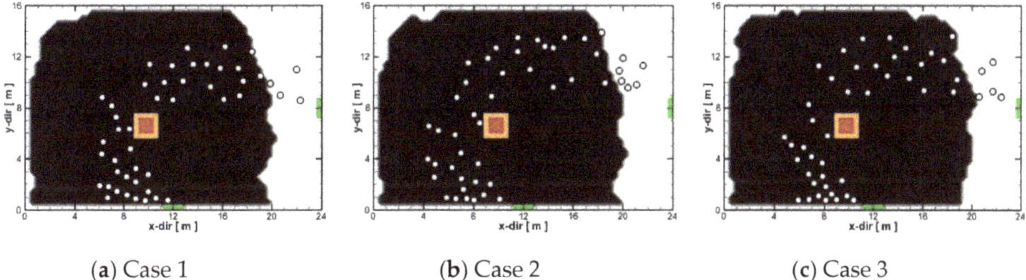

(**a**) Case 1　　　　　　　(**b**) Case 2　　　　　　　(**c**) Case 3

Figure 5. The distribution of the evacuees at 8 s.

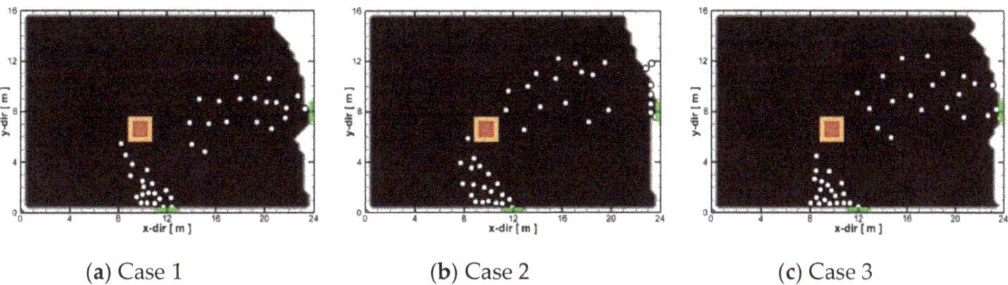

(**a**) Case 1　　　　　　　(**b**) Case 2　　　　　　　(**c**) Case 3

Figure 6. The distribution of the evacuees at 12 s.

Figure 7 represents the forces acting on the representative evacuee and the walking speed when the representative evacuee moves through the smoke region. As represented in Figure 7a, the representative evacuee for Case 1 changes its walking direction according to the influence of the motive and social forces. Additionally, in contrast with Figure 4, the magnitude of the social force rapidly increases at 17 s, which is caused by the approach of the other evacuees. Based on these acting forces, the representative evacuee rapidly decreases its walking speed because the representative evacuee moves towards the crowed space around the bottom exits. That is, it is once again confirmed that the influence of the smoke on the evacuees cannot be considered with the Helbing's movement model.

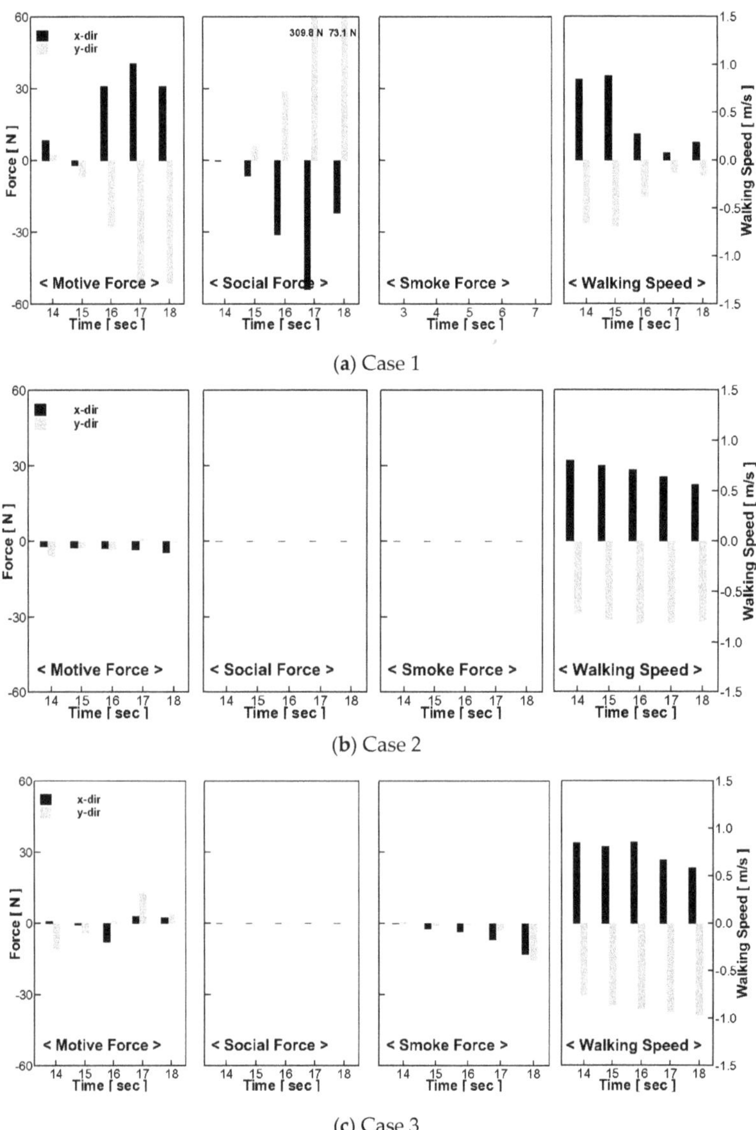

(**a**) Case 1

(**b**) Case 2

(**c**) Case 3

Figure 7. The forces acting on the representative evacuee and the walking speed from 14 to 18 s.

As illustrated in Figure 7b, the representative evacuee for Case 2 is not influenced by the inner smoke region force because it is assumed that the existing inner smoke region force is proportional to the visibility [11]. Therefore, the representative evacuee for Case 2 is only influenced by the motive force, which means that the evacuees for Case 2 move towards the exit along the predetermined desired direction. As a result of these acting forces, the evacuees for Case 2 are distributed around the center of the computational domain in the y-direction because the desired directions around the upper right side are predetermined to move towards the right side (Figure 1b).

In contrast with Figure 7b, the representative evacuee for Case 3 is influenced by the motive and smoke forces (Figure 7c). The evacuees for Case 3 are continuously influenced by the inner smoke region force when they move through the smoke region. As a result of these acting forces, the evacuees for Case 3 are distributed further away from the fire, and they maintain an extended distribution inside the smoke region (Figure 6). That is, it is confirmed that the behavioral changes of the evacuees inside the smoke region can be well considered by applying the modified BR-smoke model.

5. Conclusions

In this study, the inner smoke region force of the BR-smoke model is modified by varying the walking speed according to the smoke density. Then, the modified BR-smoke model is applied to FDS+Evac to analyze the evacuation characteristics and is compared with those of existing models (Helbing's social force model with the correlation between walking speed and smoke density and the existing BR-smoke model). The conclusions of this study are as follows:

1. It is confirmed that the behavioral changes of the evacuees due to the smoke cannot be considered with the Helbing's movement model. Thus, the BR-smoke model should be applied for considering the behavioral changes against the smoke. However, the behavioral changes inside the smoke region cannot be considered by using the existing BR-smoke model because the inner smoke region force does not act on the evacuee due to the decreased visibility.
2. The evacuation characteristics inside the smoke region can be improved by using the modified BR-smoke model because the evacuees are continuously influenced by the modified inner smoke force inside the smoke region. However, additional studies for determining a more reliable evacuee psychological factor are required to improve the reality of the modified BR-smoke model.

Author Contributions: Writing-original draft and review and editing, S.B.; Methodology, J.-H.C.; supervision, H.S.R. All authors have read and agreed to the published version of the manuscript.

Funding: This research received no external funding.

Acknowledgments: This study was supported by the research fund from Chosun University (K207896001) and Basic Science Research Program through the National Research Foundation of Korea (NRF) funded by the Ministry of Science, ICT, and Future Planning (2018R1C1B508493913).

References

1. Custer, R.L.P.; Meacham, B.J. *Introduction to Performance-Based Fire Safety. Society of Fire Protection Engineers*; NFPA: Quincy, MA, USA, 1997; ISBN 0-87765-421-2.
2. Jin, T. Visibility through Fire Smoke. *J. Fire Flammabl.* **1978**, *9*, 135–155.
3. Frantzich, H.; Nilsson, D. *Utrymning Genom tät Rök: Beteende och Förflyttning*; Report 3126; Department of Fire Safety Engineering and Systems Safety, Lund University: Lund, Sweden, 2003.
4. Purser, D.A. Toxicity Assessment of Combustion Products. In *SFPE Handbook of Fire Protection Engineering*, 2nd ed.; National Fire Protection Association: Quincy, MA, USA, 1995.
5. Chiewchengchol, W.; Koga, T.; Hirate, K. Development of an evacuation simulator using a walkthrough system and research on evacuation behavior in the case of fire. *J. Asian Archit. Build. Eng.* **2011**, *10*, 101–108. [CrossRef]
6. Fahy, R.F.; Proulx, G. Collective Common Sense: A Study of Human Behavior during the World Trade Center Evacuation. *NFPA J.* **1995**, *87*, 61–64.
7. Proulx, G. The Impact of Voice Communication Messages during a Residential Highrise Fire. In Proceedings of the First International Symposium Human Behaviour in Fire, Fire SERT Centre, University of Ulster, Belfast, UK, 30 August–2 September 1998; pp. 265–274.

8.	Proulx, G.; Fahy, R.F. Account Analysis of WTC Survivors. In Proceedings of the 3rd International Symposium on Human Behaviour in Fire 2004, Interscience Communications and University of Ulster, Belfast, UK, 1–3 September 2004; pp. 203–214.

9.	Bae, S.; Ryou, H.S. A mathematical modelling of the interaction between evacuees and fire through radiation. *Fire Technol.* **2016**, *52*, 847–864. [CrossRef]

10.	Bae, S.; Choi, J.-H.; Kim, C.; Hong, W.-H.; Ryou, H.S. Development of new evacuation model (BR-radiation model) through an experiment. *J. Mech. Sci. Technol.* **2016**, *30*, 3379–3391. [CrossRef]

11.	Bae, S.; Ryou, H.S. Development of a smoke effect model for representing the psychological pressure from the smoke. *Saf. Sci.* **2015**, *77*, 57–65. [CrossRef]

12.	Proulx, G.; Fahy, R.F. Human behavior and evacuation movement in smoke. *ASHRAE Trans.* **2008**, *114*, 159–165.

13.	Korhonen, T.; Hostikka, S. Fire dynamics simulator with evacuation. In *FDS + Evac Technical Reference and User's Guide*; VTT Technical Research Centre of Finland: Espoo, Finland, 2009.

14.	Helbing, D.; Farkas, I.; Vicsek, T. Simulating dynamical features of escape panic. *Nature* **2000**, *407*, 487–490. [CrossRef] [PubMed]

Article

Wind Tunnel Study on Wake Instability of Twin H-Rotor Vertical-Axis Turbines

Kun Wang [1], Li Zou [1,2,3,*], Aimin Wang [1], Peidong Zhao [1] and Yichen Jiang [1,*]

[1] School of Naval Architecture, Dalian University of Technology, Dalian 116024, China; kunwang@mail.dlut.edu.cn (K.W.); wam@mail.dlut.edu.cn (A.W.); zhaopeidong@mail.dlut.edu.cn (P.Z.)
[2] State Key Laboratory of Structural Analysis for Industrial Equipment, Dalian University of Technology, Dalian 116024, China
[3] Collaborative Innovation Center for Advanced Ship and Deep-Sea Exploration, Shanghai 200240, China
* Correspondence: lizou@dlut.edu.cn (L.Z.); ycjiang@dlut.edu.cn (Y.J.)

Received: 15 July 2020; Accepted: 16 August 2020; Published: 20 August 2020

Abstract: In recent years, the H-rotor vertical-axis turbine has attracted considerable attention in the field of wind and tidal power generation. After a series of complex spatiotemporal evolutions, the vortex shed from turbine blades forms a turbulent wake with a multi-scale coherent structure. An analysis of the wake characteristics of twin turbines forms the basis of array optimisation. This study aimed to examine the instability characteristics of a twin-turbine wake with two rotational configurations. The dynamic evolution characteristics of coherent structures with different scales in the wake were analysed via wavelet analysis. The results show that an inverse energy cascade process occurs after the high-frequency small-scale coherent structures induced by rotation lose their coherence. This self-organising characteristic is more apparent in the quasi two-dimensional wake of a forward-moving counter-rotating turbine (Array 1) than in that of a backward-moving counter-rotating turbine (Array 2). With greater organisation and coherence, the wake of Array 1 exhibits low-frequency instability characteristics dominated by a large-scale coherent structure. In addition, the signals reconstructed using wavelet transform show that asymmetric modes exist between low-frequency large-scale coherent structures. The experimental results provide a new perspective on the instability mechanism of twin-turbine wakes, as well as important data for numerical modelling.

Keywords: twin H-rotor vertical-axis turbines; wake; instability; wavelet transform

1. Introduction

Wind and tidal power generation technologies have advanced rapidly in recent years. Turbine wake is a type of unsteady turbulent flow. The wake effect exerts an important influence on the power generation efficiency of a turbine, fatigue life of a rotor, and stability of a power grid. Therefore, it is crucial to study the flow characteristics of turbine wake.

H-rotor turbine is the core equipment for vertical-axis wind power generation and tidal power generation [1,2]. Compared with the horizontal axis turbine, the vertical axis turbine can absorb energy in any direction of flow without complicated yaw system. The vertical axis turbine has a broad prospect in offshore wind power generation. The swept surface of a rotating vertical-axis H-rotor turbine (or horizontal axis turbine) blades is a cylinder (or disk). In many wake calculations, the turbine rotor can be described by a generalised actuator disk or actuator cylinder. To reduce the calculation for wake simulation, Rajagopalan et al. [3] used the CFD (Computational Fluid Dynamics) method to solve the generalised actuator disk rather than the dense blade grid and adopted the aerodynamic force acting on the wind turbine to the control equation. Wu [4] first described the nonlinear actuator disk model of a propeller using a formula. Although real numerical calculations were not conducted, the possibility of applying the actuator disk model to the numerical simulation of complex shapes was demonstrated. Conway [5] further studied the numerical analysis method. Madsen [6] was the first to propose a nonlinear actuator disk model for wind turbine aerodynamics and an actuator cylinder model to describe the flow field of vertical-axis wind turbines. Sørensen and Shen [7] introduced the actuator line method as an extension of the actuator disk method for non-uniform force. Shen and Zhang [8] extended the actuator line method to the actuator surface method and applied it to a vertical-axis wind turbine. The application of the method based on the actuation theory in the far wake region has yielded satisfactory results. However, research into the interaction between blade geometry and wake in this series of methods is still in progress, and the relevant theoretical knowledge requires further verification and supplementation.

Periodic vortex shedding is an instability that generally occurs in the wake of a typical blunt body, such as a cylinder or disk. This type of wake instability has also been reported in the wake of turbines. Medici and Alfredsson [9] detected a low-frequency peak in the wake spectrum. They determined it to be a manifestation of a blunt body wake mode that had evolved from the turbine's instability. Araya et al. [10] compared and analysed the wake of a vertical-axis turbine with the cylinder wake and linked them quantitatively.

Research on twin vertical-axis turbine arrays focuses mainly on the layout and output performance. Zanforlin and Nishino [11] used two-dimensional numerical simulation to study the aerodynamic performance of two counter-rotating vertical-axis turbines. It was observed that the overall power output of staggered turbines was less than that of parallel turbines. The decrease of power output depended on the direction of flow and the rotation configurations. Chen et al. [12] used a two-dimensional SST k–ω model based on DES simulation to study the power coefficients of two vertical-axis turbines by varying five factors. They observed that the five factors influenced the power factor to different levels (tip speed ratio > inflow direction angle > rotation direction > spacing > blade angle). The tip speed ratio had the greatest influence on the power output. Furthermore, the power coefficient of the two turbines was higher during reverse rotation. The power output of the two reverse rotating turbines was 9.97% higher than that of one turbine.

There are few reports on the study of the wake dynamic characteristics of twin vertical-axis turbines. Lam and Peng [13] studied the wake characteristics of a twin vertical-axis turbine with a spacing of $1D$, through wind tunnel experiments ($T/D = 2$; T = distance between the two centres; D = diameter). It was observed that the method of reverse rotation aided the wake recovery. In particular, they observed that a pair of counter-rotating vortices in the wake evolution process also contributed to the wake recovery. Subsequently, they put forward two types of turbine array arrangement. In addition, there are few reports on the instability characteristics of the wake evolution of twin vertical-axial turbines. However, there have been substantial achievements in the study of the flow around two parallel cylinders. Zdravkovich and Pridden [14] divided the flow around two parallel cylinders into three main flow patterns according to the influence of the distance ratio on the wake: single blunt body mode ($1 < T/D < 1.1$–1.2); asymmetric mode (1.1–$1.2 < T/D < 2$–2.5), and symmetric mode ($T/D > 2$–2.5). There were biased flow patterns in the asymmetric mode region (Alam et al. [15], Zhou et al. [16], Xu et al. [17]). When the fluid flowed between two parallel cylinders, a random oscillating clearance flow was formed between the two cylinders. When the clearance flow deviated to one of the cylinders, the width of the wake became narrow and the vortex shedding frequency and drag coefficient increased. Simultaneously, the wake of the other cylinder widened and the vortex shedding frequency and drag coefficient reduced.

A relationship exists between the wake of a H-rotor vertical-axis turbine and the cylinder wake. Furthermore, the two differ owing to the turbine rotation and vortex shedding by the blades. The differences and relationships between the wake characteristics of the two need to be explored further. Particularly, the wake instability, which occurs in the flow around two parallel cylinders, has rarely been reported in studies of the wake evolution of twin vertical-axial turbines. Therefore, wind tunnel tests are conducted to examine the space–time evolution characteristics of the wake of counter-rotating twin turbines with two rotational configurations, particularly the instability characteristics of the wake. The results of previous experiments [18] showed that with the increase of solidity (for example, the chord length remained unchanged at 81 mm, and the number of blades changed from three to five), the vertical axis turbine presented a low-frequency instability mode characterized by large-scale coherent structure dominating wake development. At $5D$, the five-blade turbine evolved into a low-frequency instability mode, while the wake of the three-blade turbine was still dominated by the rotation effect. In this study, wavelet analysis is also used to investigate wake instability of twin three-blade turbines.

The remainder of this paper is organised as follows: Section 2 introduces the experimental methods. Section 3.1 analyses the time-averaged characteristics of a forward-moving counter-rotating turbine (Array 1), describes the temporal and spatial evolution characteristics, and reveals the asymmetric modes between large-scale structures in the wake. Section 3.2 presents the time-averaged characteristics, evolution characteristics, and self-organisation characteristics of the wake of a backward-moving counter-rotating turbine (Array 2) and a comparative analysis with the wake of Array 1. Finally, Section 4 summarises the paper.

2. Methods

2.1. Wind Tunnel and Twin Turbine

The test section of the wind tunnel had a length of 18 m, width of 3 m, and height of 2.5 m (Figure 1: left side). The turbulence intensity at 6 m away from the entrance of the test section was 0.3% when the wind speed was 7.7 m/s. The distance between the centres of the two turbines was 1.2 times the diameter of the turbine ($T/D = 1.2$). The midpoint (O) of the line between the geometric centre points of the two turbines' cross-sections coincided with the centre point of the cross-section of the test section.

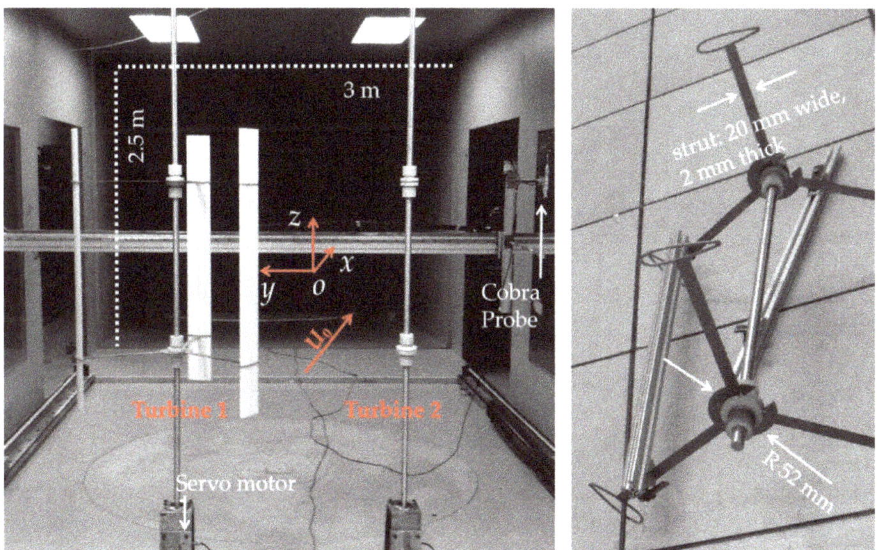

Figure 1. H-rotor turbines in the wind tunnel.

Three blades with the NACA 0018 airfoil section and a pair of 2-mm-thick struts constituted the H-shaped rotor of the turbine (Figure 1: right side). The chord length of the blade was 81 mm, and the installation angle between the blade and strut was 0°. The diameter of the turbine (D) was 600 mm, and the height was 810 mm. The two turbines were combined in different rotational directions to form two arrays (see Table 1). The direction of rotation of the turbine was observed from above.

Table 1. Description of the two arrays studied.

Array #	Turbine 1	Turbine 2
1	Anticlockwise rotation	Clockwise rotation
2	Clockwise rotation	Anticlockwise rotation

The limited size of the wind tunnel would produce a blocking effect. The following formula is provided for the solid block ratio:

$$B = \frac{\text{rotor frontal area}}{\text{test section area}}, \tag{1}$$

According to Formula (1), the solid blocking ratio was 13%, just above 10%. The blocking effect is an unavoidable problem in wind tunnel test. However, when the numerical calculation is compared with the test data, the wind tunnel measurement environment can be well reproduced [19]. Therefore, there was no blocking correction in this study. The blocking correction can refer to the correction method of Savonius vertical-axis wind turbines in reference [20]. The wake measurement area of this test was controlled to within five times of the turbine diameter (5*D*), to minimise the wake blockage effect.

2.2. Measurements

The cobra probe used for wake measurement was a series 100 model manufactured by TFI. The sampling frequency selected in this experiment was 100 Hz, which satisfied the scale requirements of the study. The measurement area is shown in Figure 2. Because the two turbines were arranged symmetrically, only the wake information of Turbine 1 was collected in the test. The sampling time of each measuring point was 30 s.

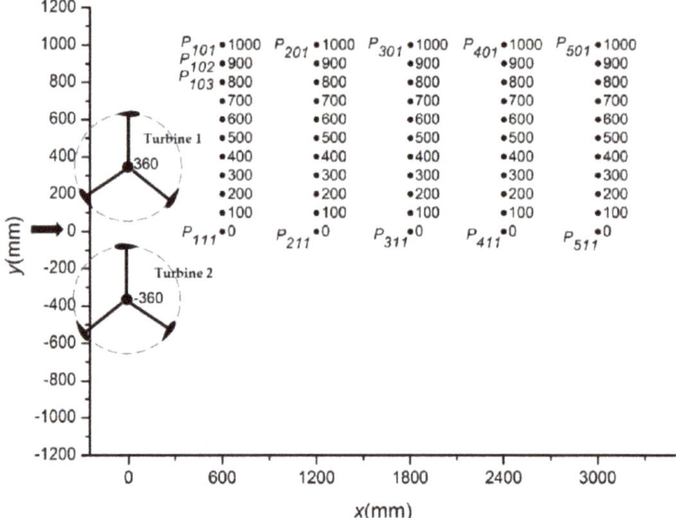

Figure 2. Measurement points at *z* = 0.

The system errors (E_S) were estimated by measuring the flow inconsistencies in the empty tunnel. The uncertainty (E_{Sx}) of the system errors in the x-direction was estimated by measuring the streamwise velocity at P_{111}, P_{211}, P_{311}, P_{411}, and P_{511}. The uncertainty (E_{Sy}) of the system errors in the y-direction was determined by collecting the streamwise component of the 21 points from P_{301} to P_{321}. The uncertainty (E_R) of the random errors was determined by repeatedly collecting the streamwise component at P_{304} in the turbine wake. The overall uncertainty was determined according to the following formula provided in reference [13]:

$$E_T = \sqrt{(E_R)^2 + (E_{Sx})^2 + (E_{Sy})^2},$$
(2)

The measurements yielded the following: $E_{Sx} = 0.9\%$, $E_{Sy} = 1.2\%$, and $E_R = 0.4\%$. Thus, the overall uncertainty was 1.6%.

Moreover, the turbine torque output at different rotation speed was measured. The torque output curves are given in Figure A1 of the Appendix A. The wake of a turbine depends on its operating state [21]. The operating state at 500 rpm was selected as a typical condition to collect wake information.

2.3. Wavelet Analysis Method

The wavelet coefficient formula of a signal *f(t)* can be expressed as

$$T(a,b) = (f(t), \psi_{a,b}(t)) = \frac{1}{\sqrt{a}} \int_{-\infty}^{+\infty} f(t)\psi^*(\frac{t-b}{a})dt,$$
(3)

T(a, b) is called the wavelet transform coefficient, and *t* is the time variable. Here, *a* is the expansion variable, and *b* is the translation variable. Ψ is the mother wavelet, and Ψ^* is the conjugate of Ψ. A Morlet wavelet [22] was used as the mother wavelet in this study. Both wavelet and Fourier are integral transforms. Expanding a function under a wavelet basis function implies the projection of a function onto a two-dimensional time-scale plane.

Wavelet analysis is an important method for turbulence research. It compensates for the spectrum analysis' dependence on the number of samples and time interval. It can be used for a localisation analysis. In a wavelet analysis, the Fourier transform basis function is replaced with a 'wavelet function'. It has a good time-frequency analysis function, can conduct multi-scale and multi-resolution analyses of time series signals, and is an effective tool for analysing the energy distribution of various frequency component signals and for extracting local information. Moreover, the variation in the wavelet energy spectrum intensity with time and frequency can reveal other physical mechanisms [23].

3. Results and Discussion

3.1. Forward-Moving Counter-Rotating Turbine

For Array 1, Turbine 1 rotates anticlockwise, and Turbine 2 rotates clockwise (Figure 3b). The blades passing through the intermediate clearance between the two turbines always move forward (x +) in the wind direction.

3.1.1. Time-Averaged Wake Characteristics

Figure 3a shows the normalised time-averaged streamwise velocity distribution. Figure 3b is the interpolated cloud chart of Figure 3a. In this test, the free inflow (U_0) was maintained at 7.7 m/s. Both the turbines rotated at 500 rpm. Figure 3 shows that there is a velocity deficit region downstream of the rotating rotor, owing to momentum extraction and the blocking effect. The speed is not immediately reduced to the maximum attenuation value when the free inflow passes through the turbine rotor. It first moves downstream for a certain distance, attains the maximum speed attenuation value, and then recovers gradually. The interaction between the clearance flow and wake flow results in the development of the intermediate shear layer, whereas the interaction between the free inflow and wake flow results in the development of the lateral shear layer. Figure 3 shows that the shear layer expands continuously in the process of developing downstream. In addition, wake excursions induced by the rotor's rotation can be observed. The above characteristics are in accordance with the literature [24–26].

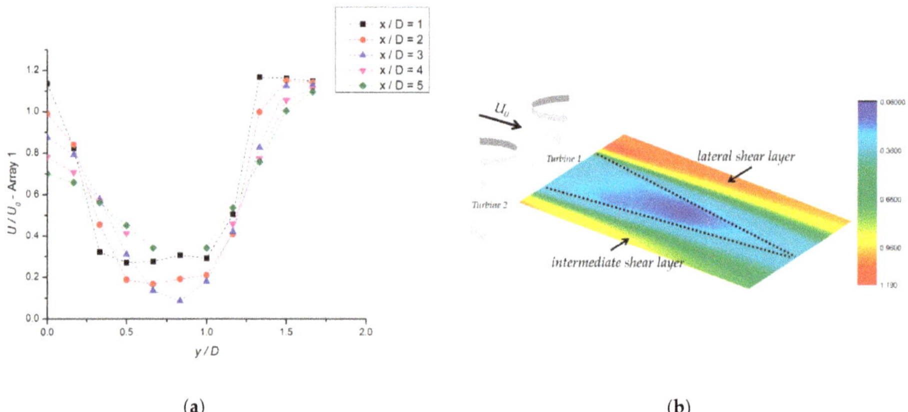

(a)   (b)

Figure 3. Time-averaged streamwise velocity distribution (**a**) and interpolation nephogram (**b**) of wake of Array 1.

In particular, the velocity information of a few measuring points at $4D$ and $5D$ in Figure 3a is absent. There are a large number of zero values in the velocity time series measured at $4D$ and $5D$. The time series defaults to zero because the cobra probe cannot capture the return velocity. Although the probe can output the time-averaged velocity information, it cannot reflect the flow fact. Therefore, the velocity information of the re-circulation region at $4D$ and $5D$ in Figure 3a is not analysed statistically. The re-circulation region had been addressed in accordance with reference [27]. Notably, the negative velocity information of the re-circulation region is not reflected in Figure 3b. This is because the cloud images in this paper are interpolation cloud images. The cloud charts in this paper are only for illustration.

Figure 4 shows the distribution of normalised normal Reynolds stress $\overline{uu}/U_0^2$ and the interpolated cloud chart. The maximum value appears at $1D$ of the intermediate shear layer, where a strong shear is formed between the clearance flow and wake flow. The secondary maximum value appears in the lateral shear layer downstream of the re-circulation region. The maximum values are concentrated in the shear layer with a sharp velocity gradient. The shear layer contains coherent structures of different scales. These coherent structures develop and brake, which causes the low-speed fluid in the wake to mix with the high-speed free inflow outside. Thereby, the momentum of the external free inflow is transferred to the wake, which causes the wake area and shear layer to expand gradually. Next, the wavelet transform is used to determine the dynamic evolution characteristics.

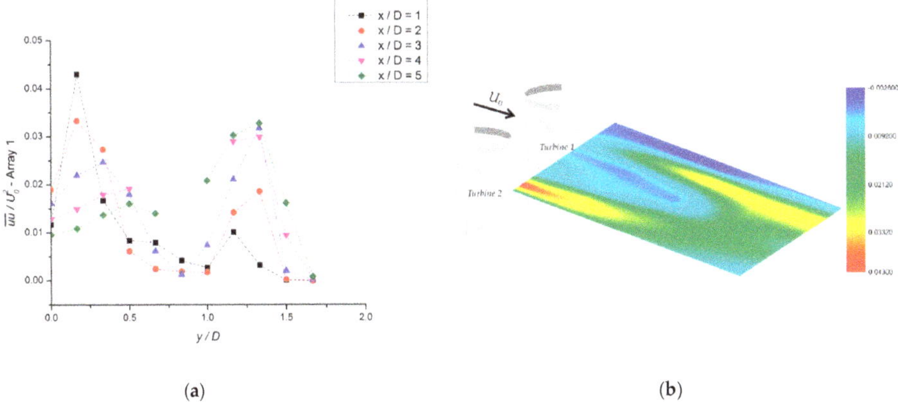

(a)   (b)

Figure 4. Normal Reynolds stress distribution (**a**) and interpolation nephogram (**b**) of wake of Array 1.

3.1.2. Dynamic Characteristics of Evolution

Figure 5b–d shows the wavelet energy spectrum of the measuring points in the lateral shear layer. Blade vortex shedding ($3f$) dominates the wake development at P_{103} (Figure 5b). In addition, a coherent structure with a lower frequency appears, which indicates that the blade vortex shedding begins to lose coherence gradually when the wake reaches $1D$. As the wake continues to develop downstream and reaches $3D$ P_{303} (Figure 5c), the blade vortex shedding breaks completely and disappears. However, the coherent structure induced by rotation still dominates the wake development. The coherent structure with the rotating frequency f as the dominant frequency can be observed clearly. Meanwhile, the energy of the low-frequency coherent structure continues to increase, whereas that of the high-frequency coherent structure continues to decrease. When the wake coherent structure develops to $4D$ P_{402} (Figure 5d), the high-frequency small-scale coherent structure induced by rotor rotation breaks and disappears. Furthermore, the low-frequency coherent structures are relatively stable and independent, thereby dominating the wake development. The low-frequency coherent structure of $4D$ is the result of the evolution of the coherent structure of $3D$. In the evolution process, the dominant frequency of the coherent structure decreases gradually, and its scale continues to increase.

Figure 6b,c is the wavelet energy spectrum of the measuring points in the intermediate shear layer. The blade vortex shedding has broken completely and disappeared at $1D$ P_{110}, unlike in the case of the lateral shear layer. The small-scale coherent structure with frequency f induced by rotation dominates the wake development. A low-frequency large-scale coherent structure appears simultaneously. At $2D$ P_{210}, the coherence of the low-frequency large-scale structures increases gradually, whereas most of the small-scale coherent structures induced by rotation have broken.

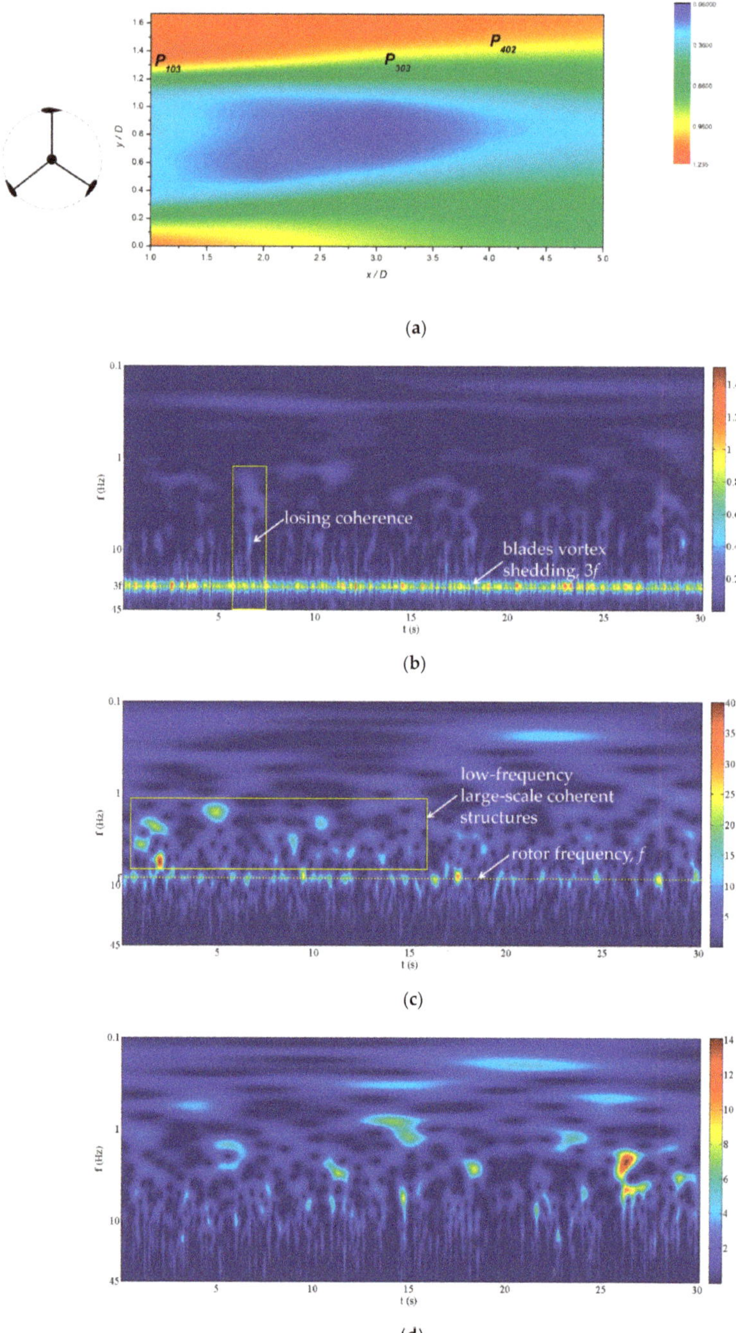

Figure 5. Wavelet transform analysis of time series at different measuring points in the lateral shear layer: (**a**) schematic diagram, (**b**) P_{103}, (**c**) P_{303}, (**d**) P_{402}.

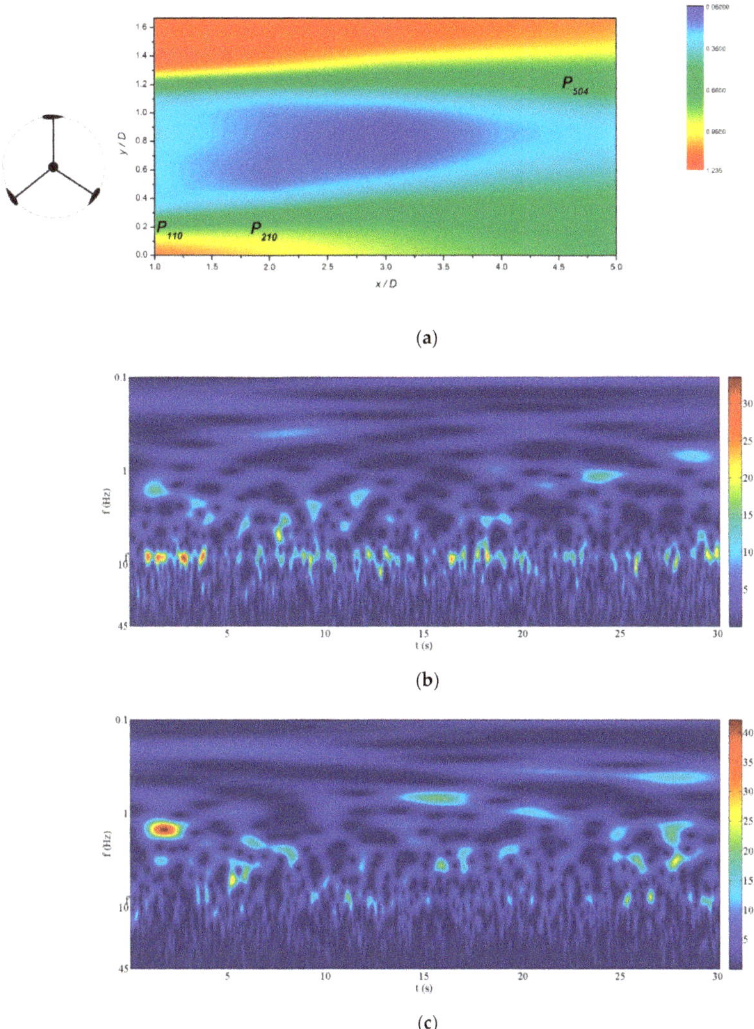

Figure 6. Wavelet transform analysis of time series at different measuring points in the intermediate shear layer: (**a**) schematic diagram, (**b**) P_{110}, (**c**) P_{210}.

The direction of blade movement through the two turbine intervals of Array 1 is identical to that of the incoming flow. Furthermore, the direction of blade movement on both the sides is opposite to that of the incoming flow. This rotational configuration promotes the development of clearance flow, which accelerates the loss of coherence of blade vortex shedding and the occurrence of shear layer instability. This rotational configuration also causes the wakes of the two turbines to shift outwards relative to each other. A strong shear is produced by the outward offset wake and the free inflow. In the lateral shear layer, the blade vortex shedding loses coherence in $2D$ under strong shear. As the blade vortex shedding loses coherence, the rotation induction effect disappears gradually and the energy of the high-frequency small-scale coherent structure decreases gradually. Then, the dominant frequencies ($3D$ and $4D$) of the wake coherent structure gradually decrease and the scale gradually increases. The time-averaged wake velocity distribution (Figure 3a) indicates the presence of a re-circulation region at $4D$ and $5D$. In addition to the re-circulation region, $5D$ P_{504} (Figure 6a) is selected, and the wavelet energy spectrum of its time series signal is shown in Figure 7. As is evident from Figure 7, as the wake development proceeds, the organisation of the coherent structure is enhanced, and the energy is concentrated in the low-frequency range of 2–3 Hz. The constant Strouhal number is between 0.16 and 0.23 with respect to D. The low-frequency instability characteristic of the large-scale coherent structure dominating the wake development was mentioned in reference [10]. Therein, the wake of a vertical-axis turbine evolved into a blunt body wake mode, and the Strouhal number was approximately 0.26. However, the Strouhal number in this study fluctuates in a narrow frequency band.

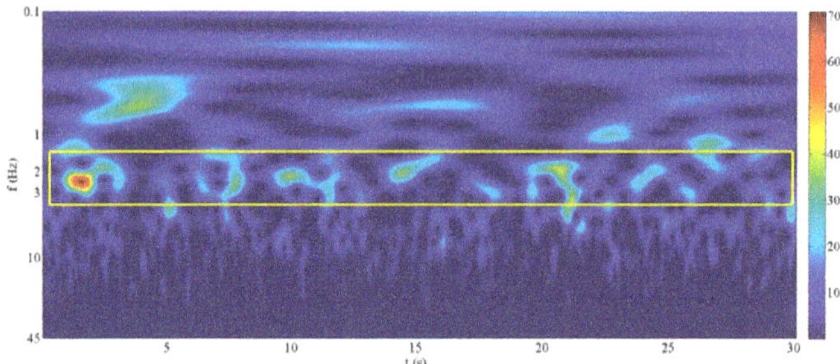

Figure 7. Wavelet transform analysis of time series at P_{504}.

As mentioned earlier, the rotating configuration of Array 1 causes the two turbines' wakes to diverge away from each other. This promotes the development of clearance flow. It can be inferred that the clearance flow that is biased towards a turbine causes the wake on this side to narrow with an increase in the shedding frequency. Simultaneously, the clearance flow causes the wake on the other side to widen with a decrease in the shedding frequency. That is, there are asymmetric modes between the large-scale coherent structures in the wake evolution process of the twin turbines in Array 1. Next, wavelet decomposition and reconstruction are applied to demonstrate this flow pattern.

3.1.3. Asymmetrical Mode

Figure 8 is a local enlarged image of Figure 7, which is a typical wavelet energy spectrum of the scale of dynamic variation, for the period 13–19 s. This figure also reflects the intermittent process of turbulence effectively. The top of Figure 9 is the original time series signal of P_{504}, and the frequency range of the original signal is 0–48.4 Hz. The bottom graph in Figure 9 shows the fifth layer high frequency component u_5 after wavelet decomposition and reconstruction of the original signal; u_5 has a frequency range of 1.5–3 Hz. In this study, the db3 wavelet is used for the decomposition and reconstruction.

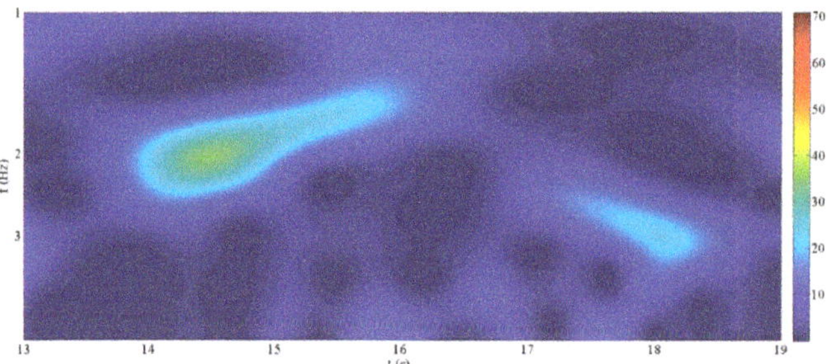

Figure 8. The 13–19 s local graph of Figure 7.

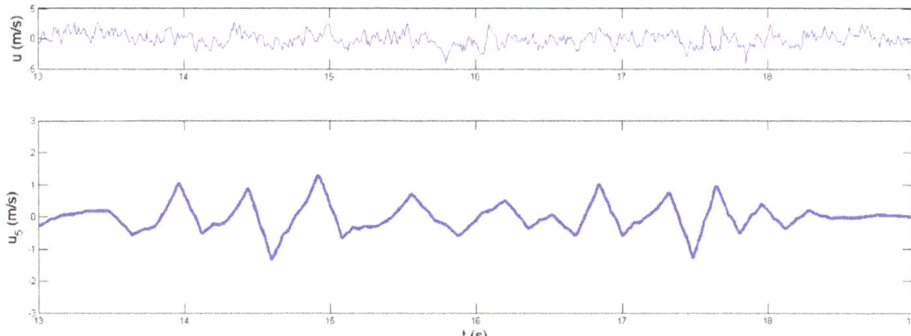

Figure 9. Original time series signal of P_{504} (**top**) and the reconstructed signal (**bottom**).

As is evident from Figure 8, the frequency of the coherent structure is mainly about 2 Hz between 14 and 15 s. Furthermore, it is in a state of continuous reduction. The dominant frequency of the coherent structure continues to decrease in 15–16 s and attains 1 Hz. That is, in 14–16 s, the coherent structure is in a mode where the dominant frequency decreases continuously. The continuous decrease in the dominant frequency implies that the scale of the coherent structure continues to increase, i.e., the wake continues to widen. The reconstructed velocity time series (Figure 9: bottom) also shows the dynamic process of the gradually increasing cycle in 14–16 s. The periodicity disappears in 16–17 s. Another mode appears in the coherent structure in 17–18 s (Figure 8). The dominant frequency of the coherent structure begins to increase, and the scale decreases continuously, i.e., the wake continues to narrow.

The previous evolution process reveals that the low-frequency large-scale coherent structure dominates the wake development of Array 1. In addition, the low-frequency instability fluctuates in a narrow frequency band. Asymmetric modes exist between the large-scale coherent structures.

3.2. Backward-Moving Counter-Rotating Turbine

For Array 2, Turbine 1 rotates clockwise, and Turbine 2 rotates anticlockwise (Figure 10b). The direction of blade movement through the intermediate clearance of the two turbines is always backward (x is negative) against the wind.

3.2.1. Time-Averaged Wake Characteristics

As shown in Figure 10a,b, the degree of expansion of the intermediate shear layer in the wake of Array 2 is less than that of Array 1. This is because the two turbines' wakes in Array 2 are close to each other. The Reynolds normal stress in the intermediate shear layer (Figure 10c,d) first increases and then decreases.

The normal stress distribution of the lateral shear layer (Figure 10c,d) has the maximum value at $1D$. Thereafter, it shows a weakening trend. This is because Turbine 1 rotates clockwise and the direction of blade movement through the lateral shear layer is identical to that of the incoming flow. The normal stress distribution trend of the lateral shear layer of Array 1 (Figure 4a,b) is contrary to that of Array 2 owing to the different rotational configurations.

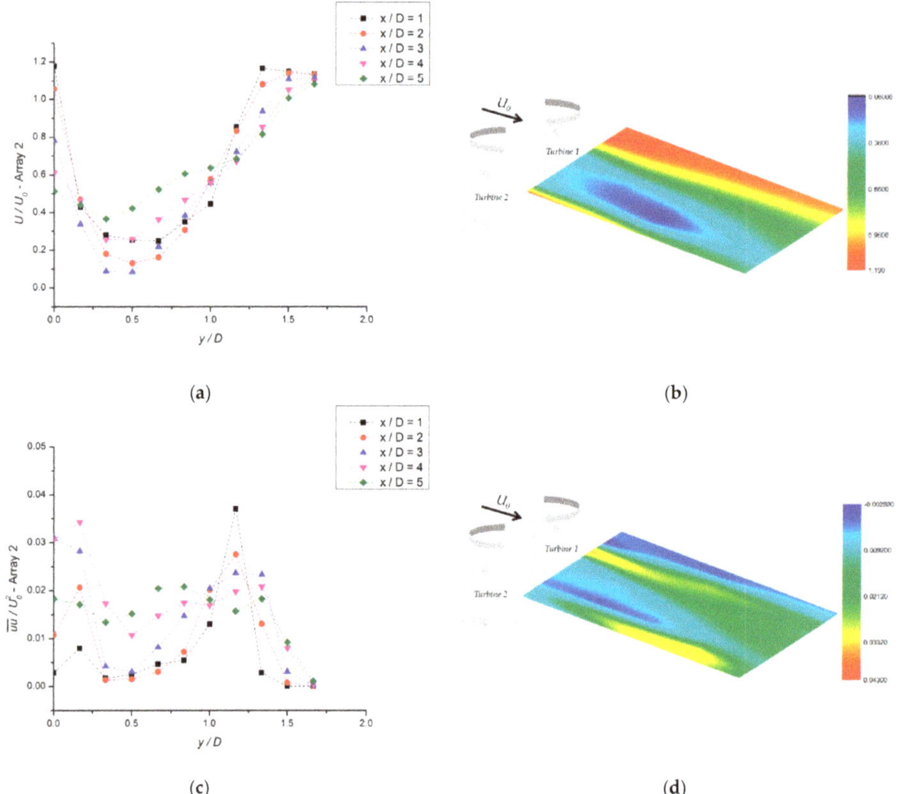

Figure 10. Time-averaged streamwise velocity distribution (**a**) and interpolation nephogram (**b**) of wake of Array 2. Normal Reynolds stress distribution (**c**) and interpolation nephogram (**d**) of wake of Array 2.

3.2.2. Dynamic Characteristics of Evolution

In the intermediate shear layer $1D\ P_{110}$ (Figure 11b) of the wake of Array 2, the blade vortex shedding interaction increases. However, the wake development continues to be dominated by the blade vortex shedding with a frequency of $3f$. However, the blade vortex shedding of Array 1 completely disappears at $1D\ P_{110}$ (Figure 6b). At $3D\ P_{311}$ (Figure 11c), the rotation induction effect still exists. While the wake of Array 1 is in $2D$ (Figure 6c), most of the coherent structures with frequency f have broken. At $5D\ P_{503}$ (Figure 11d), although the energy of the low-frequency large-scale coherent structures has increased, their energy is not concentrated, and their organisation and coherence are not strong. In addition, the high-frequency small-scale coherent structure induced by rotation still exists.

It is evident from the evolution process of the coherent structure that the interaction between the lateral shear and free inflow is weaker than that of Array 1 because the wakes of the two turbines in Array 2 are close to each other. The high-frequency small-scale coherent structure passing through the lateral shear layer loses coherence after that of Array 1. This rotation configuration also hinders the development of clearance flow. The high-frequency small-scale coherent structure induced by rotation passing through the clearance also loses coherence after that of Array 1. In conclusion, the Array 2 turbine does not exhibit the low-frequency instability characteristic of the large-scale coherent structure dominating the wake development.

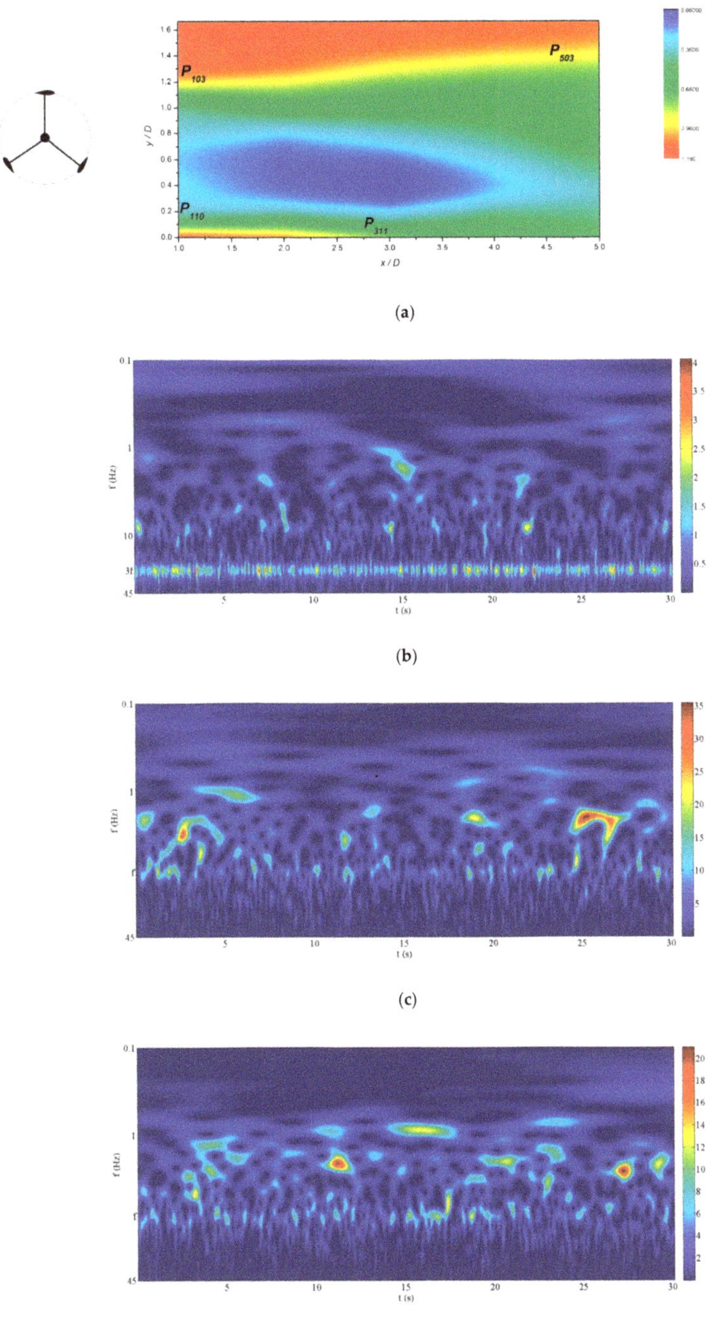

Figure 11. Wavelet transform analysis of time series at different measuring points in the shear layer: (**a**) schematic diagram, (**b**) P_{110}, (**c**) P_{311}, (**d**) P_{503}.

3.2.3. Self-organisation Characteristics

Previous research has shown that the high-frequency small-scale coherent structures (frequency $3f$ and f) induced by rotation dominate the wake development in the near wake. As shown in Figure 12b (local part of Figure 12a), at $1D$ P_{103} of the lateral shear layer of Array 2, the coherent structure with the frequency of $3f$ gradually loses coherence and disintegrates. Thereby, it draws energy from the average flow and transfers it to the smaller scale vortex structure (down arrow in Figure 12b). This process is consistent with the energy cascade phenomenon. In addition, the vortex structures merge, the frequency decreases, and the scale increases (the up arrow in Figure 12b).

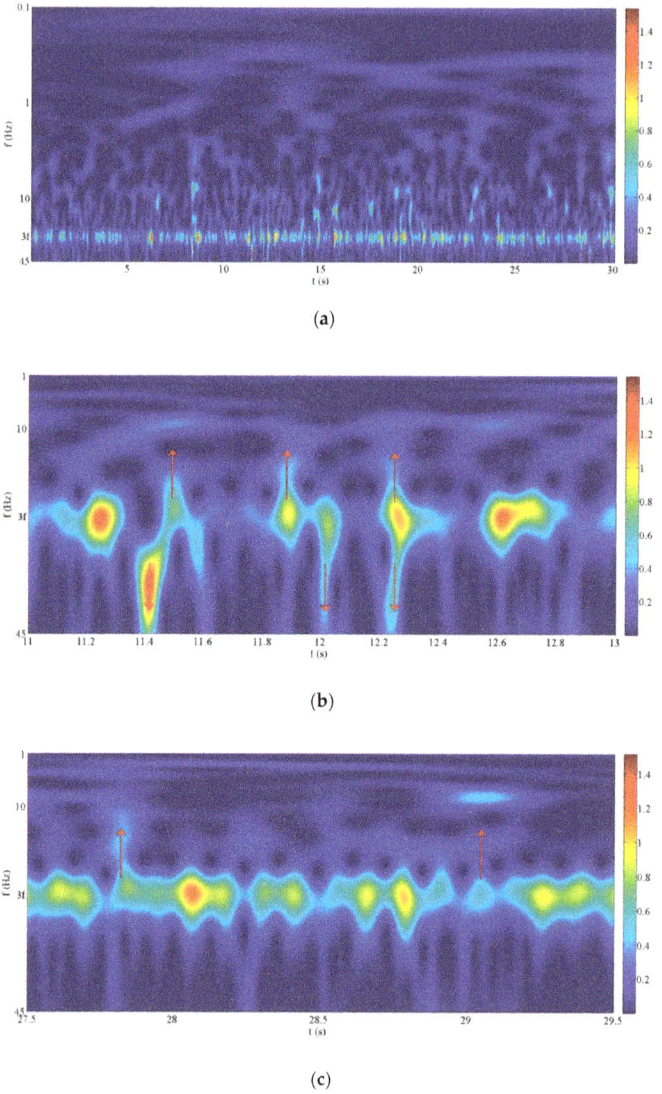

Figure 12. Wavelet transform analysis of time series at P_{103} (**a**). Local parts of Figure 12a (**b**) and Figure 5b (**c**).

At $1D$ P_{103} of the lateral shear layer of Array 1, the adjacent small-scale coherent structures interact with each other, the dominant frequency decreases gradually, and the scale increases gradually (the up arrow in Figure 12c). It is evident from the previous research that, as the wake continues to develop downstream, the organisation and coherence of large-scale coherent structures become stronger. Furthermore, the energy is gradually transferred to the low-frequency structure. Finally, at $5D$ P_{504} (Figure 7), the background small-scale coherent structure is very weak.

The above evolution process illustrates that there is an inverse energy cascade process in the wakes of Array 1 and Array 2 and that the inverse energy cascade process of Array 1 is more apparent. In [28], it was indicated for the first time that there is a series of inverse energy cascade processes in two-dimensional isotropic turbulence, i.e., the energy transferred from a small- to a large-scale structure in the development process of two-dimensional turbulence. Subsequently, many scholars studied two-dimensional turbulence by numerical simulation or experimental methods and further verified the existence of two-dimensional turbulent inverse energy cascade. It was observed that the adjacent small-scale vortices in the flow field interacted with each other and merged into larger vortices. Finally, one or several relatively stable self-organised large vortices with the same size as that of the physical domain could be formed. These are the self-organisation characteristics of two-dimensional turbulence [29–34]. In essence, turbulence is a completely three-dimensional nonlinear fluid motion. However, certain specific flows display apparent quasi two-dimensional characteristics [28]. This is mainly because the scale of such a flow in one direction is significantly smaller than that in the other two directions.

Figure 13 shows the profiles of the six components of Reynolds stress at $1D$ and $5D$ for the wakes of Array 1 and Array 2. The three components of Reynolds normal stress ($\overline{uu}$, $\overline{vv}$, $\overline{ww}$) in the wake of Array 1 (Figure 13a) are almost identical at $1D$, whereas the Reynolds shear stress component ($\overline{uw}$, $\overline{vw}$) is essentially zero. For Array 2 (Figure 13b), the streamwise direction and vertical direction in the $1D$ lateral shear layer have an apparent correlation ($\overline{uw}$), and a shear effect is observable. That is, the three-dimensional characteristics of the wake of Array 2 are more apparent than those of Array 1 at $1D$. This is true at $5D$ as well (Figure 13c,d).

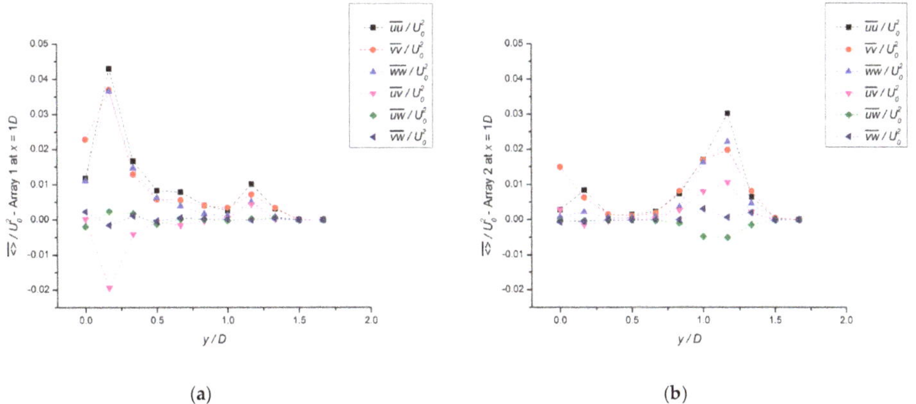

(a)

(b)

Figure 13. *Cont.*

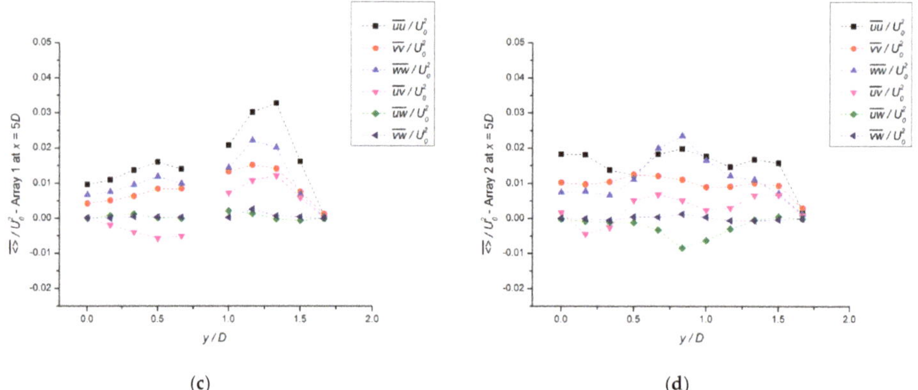

Figure 13. The profiles of the six components of Reynolds stress at $1D$ (**a**,**b**) and $5D$ (**c**,**d**) for the wakes of Array 1 and Array 2.

In the quasi two-dimensional wake of Array 1, the small-scale structure continuously transfers energy to the large-scale structure in accordance with the principle of inverse energy cascade. Therefore, the wake flow is constantly modified to carry more energy. However, in the three-dimensional wake of Array 2, the coherent structure continuously absorbs energy from the free inflow to transfer it to the smaller turbulence structure. Finally, the energy is dissipated from the smallest turbulence vortex and is converted to the internal energy of the fluid. As a result, the time-averaged streamwise velocity profile of Array 2 (Figure 10a) is flatter than that of Array 1 (Figure 3a).

4. Conclusions

This paper focuses on the wake instability characteristics of two types of counter-rotating twin turbines with a spacing of $0.2D$ ($T/D = 1.2$). The rotational configuration of Array 1 (forward-moving counter-rotating turbine) causes the formation of strong shear in the lateral shear layer and promotes the development of clearance flow. This, in turn, results in the formation of strong shear in the intermediate shear layer. The high-frequency small-scale coherent structure in the near wake induced by rotation gradually loses coherence in the intermediate and lateral shear layers. This causes the shear layers to become unstable. The instability mechanism is induced earlier in the wake of Array 1 than it is in the wake of Array 2 (backward-moving counter-rotating turbine). After the high-frequency small-scale coherent structure loses coherence, energy is transferred to the smaller-scale coherent structure according to the energy cascade principle. Meanwhile, the small-scale coherent structure transfers energy to the low-frequency coherent structure to form a self-organised large-scale coherent structure. This type of self-organisation characteristics is more apparent in the quasi two-dimensional wake of Array 1 than that in the wake of Array 2. With the development of the wake downstream, the organisation, coherence, and scale of the low-frequency structure in the wake of Array 1 is enhanced gradually. Ultimately, low-frequency large-scale coherent structures dominate the wake development. In addition, there are asymmetric modes between the large-scale coherent structures in the wake of Array 1, which causes the Strouhal number to fluctuate between 0.16 and 0.23.

These results have certain reference significance for studying wake flow mechanism and optimizing turbine array layout. The self-organised large-scale coherent structure has a positive role in promoting turbulent mixing and entrainment, but also has a negative effect on fatigue load and power fluctuation. Further studies can consider the flow control of wake large-scale coherent structure.

Author Contributions: Conceptualization, K.W.; data curation, K.W.; formal analysis, K.W.; investigation, K.W.; methodology, K.W. and A.W.; project administration, Y.J.; resources, L.Z.; software, A.W.; supervision, Y.J.;

validation, A.W.; visualization, K.W.; writing—original draft, K.W.; writing—review and editing, Y.J. and P.Z. All authors have read and agreed to the published version of the manuscript.

Funding: This research was funded by Qingdao National Laboratory for Marine Science and Technology (QNLM2016ORP0402); Defense Industrial Technology Development Program (SXJQR2018WDKT02); State Key Laboratory of Structural Analysis for Industrial Equipment(S18408); and National Natural Science Foundation of China (51975032) & (51939003).

Acknowledgments: The authors would like to thank Qingdao National Laboratory for Marine Science and Technology.

Conflicts of Interest: The authors declare no conflict of interest.

Appendix A

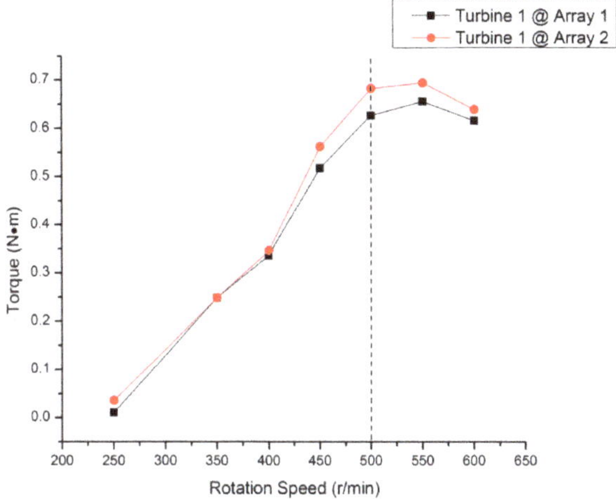

Figure A1. Torque output curves at different rotation speeds.

References

1. Balduzzi, F.; Bianchini, A.; Carnevale, E.A.; Ferrari, L.; Magnani, S. Feasibility analysis of a Darrieus vertical-axis wind turbine installation in the rooftop of a building. *Appl. Energy* **2012**, *97*, 921–929. [CrossRef]
2. Tjiu, W.; Marnoto, T.; Mat, S.; Ruslan, M.H.; Sopian, K. Darrieus vertical axis wind turbine for power generation ii: Challenges in HAWT and the opportunity of multi-megawatt darrieus VAWT development. *Renew. Energy* **2015**, *75*, 560–571. [CrossRef]
3. Rajaopalan, R.G.; Rickerl, T.L.; Klimas, P.C. Aerodynamic interference of vertical axis wind turbines. *J. Propul. Power* **1990**, *6*, 645–653. [CrossRef]
4. Wu, T.Y. Flow through a heavily loaded actuator disc. *Schiffstechnik* **1962**, *47*, 134–138.
5. Conway, J.T. Exact actuator disk solutions for non-uniform heavy loading and slipstream contraction. *J. Fluid Mech.* **2000**, *365*, 235–267. [CrossRef]
6. Madsen, H.A. The Actuator Cylinder—A Flow Model for Vertical Axis Wind Turbines. Ph.D. Thesis, Aalborg University Centre, Aalborg, Denmark, 1982.
7. Sørensen, J.N.; Shen, W.Z. Numerical modeling of wind turbine wakes. *J. Fluids Eng. Trans. ASME* **2002**, *124*, 393–399. [CrossRef]
8. Shen, W.Z.; Zhang, J.H.; Sørensen, J.N. The actuator surface model: A new navier-stokes based model for rotor computations. *J. Sol. Energy Eng. Trans. ASME* **2009**, *131*, 110021–110029. [CrossRef]
9. Medici, D.; Alfredsson, P.H. Measurements on a wind turbine wake: 3D effects and bluff body vortex shedding. *Wind Energy* **2006**, *9*, 219–236. [CrossRef]
10. Araya, D.B.; Colonius, T.; Dabiri, J.O. Transition to bluff-body dynamics in the wake of vertical-axis wind turbines. *J. Fluid Mech.* **2017**, *813*, 346–381. [CrossRef]

11. Zanforlin, S.; Nishino, T. Fluid dynamic mechanisms of enhanced power generation by closely spaced vertical axis wind turbines. *Renew. Energy* **2016**, *99*, 1213–1226. [CrossRef]

12. Chen, W.; Chen, C.; Huang, C.; Hwang, C. Power output analysis and optimization of two straight-bladed vertical-axis wind turbines. *Appl. Energy* **2017**, *185*, 223–232. [CrossRef]

13. Lam, H.F.; Peng, H.Y. Measurements of the wake characteristics of co- and counter-rotating twin h-rotor vertical axis wind turbines. *Energy* **2017**, *131*, 13–26. [CrossRef]

14. Zdravkovich, M.M.; Pridden, D.L. Interference between two circular cylinders; Series of unexpected discontinuities. *J. Ind. Aerodyn.* **1977**, *2*, 255–270. [CrossRef]

15. Alam, M.M.; Moriya, M.; Sakamoto, H. Aerodynamic characteristics of two side-by-side circular cylinders and application of wavelet analysis on the switching phenomenon. *J. Fluids Struct.* **2003**, *18*, 325–346. [CrossRef]

16. Zhou, Y.; Zhang, H.J.; Yiu, M.W. The turbulent wake of two side-by-side circular cylinders. *J. Fluid Mech.* **2002**, *458*, 303–332. [CrossRef]

17. Xu, S.J.; Zhou, Y.; So, R.M.C. Reynolds number effects on the flow structure behind two side-by-side cylinders. *Phys. Fluids* **2003**, *15*, 1214–1219. [CrossRef]

18. Zou, L.; Wang, K.; Jiang, Y.; Wang, A.; Sun, T. Wind tunnel test on the effect of solidity on near wake instability of vertical-axis wind turbine. *J. Mar. Sci. Eng.* **2020**, *8*, 365. [CrossRef]

19. Battisti, L.; Persico, G.; Dossena, V.; Paradiso, B.; Raciti Castelli, M.; Brighenti, A.; Benini, E. Experimental benchmark data for H-shaped and troposkien VAWT architectures. *Renew. Energy* **2018**, *125*, 425–444. [CrossRef]

20. Ross, I.; Altman, A. Wind tunnel blockage corrections: Review and application to savonius vertical-axis wind turbines. *J. Wind. Eng. Ind Aerod.* **2011**, *99*, 523–538. [CrossRef]

21. Posa, A. Influence of Tip Speed Ratio on wake features of a Vertical Axis Wind Turbine. *J. Wind. Eng. Ind. Aerod.* **2020**, *197*, 104076. [CrossRef]

22. Torrence, C.; Compo, G.P. A practical guide to wavelet analysis. *Bull. Am. Meteorol. Soc.* **1998**, *79*, 61–78. [CrossRef]

23. Farge, M. Wavelet transforms and their applications to turbulence. *Ann. Rev. Fluid Mech.* **1992**, *24*, 395–457. [CrossRef]

24. Howell, R.; Qin, N.; Edwards, J.; Durrani, N. Wind tunnel and numerical study of a small vertical axis wind turbine. *Renew. Energy* **2010**, *35*, 412–422. [CrossRef]

25. Tescione, G.; Ragni, D.; He, C.; Ferreira, C.J.S.; van Bussel, G.J.W. Near wake flow analysis of a vertical axis wind turbine by stereoscopic particle image velocimetry. *Renew. Energy* **2014**, *70*, 47–61. [CrossRef]

26. Posa, A.; Parker, C.M.; Leftwich, M.C.; Balaras, E. Wake structure of a single vertical axis wind turbine. *Int. J. Heat Fluid Flow* **2016**, *61*, 75–84. [CrossRef]

27. Sumner, D.; Heseltine, J.L.; Dansereau, O.J.P. Wake structure of a finite circular cylinder of small aspect ratio. *Exp. Fluids* **2004**, *37*, 720–730. [CrossRef]

28. Kraichnan, R.H. Inertial ranges in two-dimensional turbulence. *Phys. Fluids* **1967**, *10*, 1417–1423. [CrossRef]

29. Maassen, S.R.; Clercx, H.J.H.; van Heijst, G.J.F. Self-organization of quasi-two-dimensional turbulence in stratified fluids in square and circular containers. *Phys. Fluids* **2002**, *14*, 2150–2169. [CrossRef]

30. Schneider, K.; Farge, M. Decaying two-dimensional turbulence in a circular container. *Phys. Rev. Lett.* **2005**, *95*, 244502. [CrossRef]

31. Clercx, H.J.H.; Maassen, S.R.; van Heijst, G.J.F. Spontaneous spin-up during the decay of 2D turbulence in a square container with rigid boundaries. *Phys. Rev. Lett.* **1998**, *80*, 5129–5132. [CrossRef]

32. Molenaar, D.; Clercx, H.J.H.; Van Heijst, G.J.F. Transition to chaos in a confined two-dimensional fluid flow. *Phys. Rev. Lett.* **2005**, *95*, 104503. [CrossRef] [PubMed]

33. Frisch, U.; Sulem, P.L. Numerrical simulation of the inverse cascade in two-dimensional turbulence. *Phys. Fluids* **1984**, *27*, 1921–1923. [CrossRef]

34. Qian, J. Inverse energy cascade in two-dimensional turbulence. *Phys. Fluids* **1986**, *29*, 3608–3611. [CrossRef]

Article

Multiphysics CFD Simulation for Design and Analysis of Thermoelectric Power Generation

Olle Högblom and Ronnie Andersson *

Department of Chemistry and Chemical Engineering, Chalmers University of Technology,
SE-41296 Gothenburg, Sweden; olle.hogblom@pulpac.com
* Correspondence: ronnie.andersson@chalmers.se; Tel.: +46-317-722-941

Received: 30 July 2020; Accepted: 20 August 2020; Published: 22 August 2020

Abstract: The multiphysics simulation methodology presented in this paper permits extension of computational fluid dynamics (CFD) simulations to account for electric power generation and its effect on the energy transport, the Seebeck voltage, the electrical currents in thermoelectric systems. The energy transport through Fourier, Peltier, Thomson and Joule mechanisms as a function of temperature and electrical current, and the electrical connection between thermoelectric modules, is modeled using subgrid CFD models which make the approach computational efficient and generic. This also provides a solution to the scale separation problem that arise in CFD analysis of thermoelectric heat exchangers and allows the thermoelectric models to be fully coupled with the energy transport in the CFD analysis. Model validation includes measurement of the relevant fluid dynamic properties (pressure and temperature distribution) and electric properties (current and voltage) for a turbulent flow inside a thermoelectric heat exchanger designed for automotive applications. Predictions of pressure and temperature drop in the system are accurate and the error in predicted current and voltage is less than 1.5% at all exhaust gas flow rates and temperatures studied which is considered very good. Simulation results confirm high computational efficiency and stable simulations with low increase in computational time compared to standard CFD heat-transfer simulations. Analysis of the results also reveals that even at the lowest heat transfer rate studied it is required to use a full two way coupling in the energy transport to accurately predict the electric power generation.

Keywords: computational fluid dynamics (CFD), multiphysics; heat transfer; thermoelectricity; automotive

---

1. Introduction

Over the last few decades there has been a growing interest in using thermoelectric technology to increase the energy efficiency of various heat recovery systems. The applications range from stationary systems e.g., heat recovery from solar [1–3], geothermal [4,5], to heat recovery from mobile heat sources such as automotive applications [6–10] and from nuclear sources for space exploration purposes [11]. Research is also directed towards cooling strategies and optimization methods to increase the conversion efficiency in the thermoelectric systems [12]. The emission legislations in the automotive industry together with the energy prices makes it more profitable—and in some cases necessary with energy recuperation. Inside the exhaust gas recirculation (EGR) cooler in a diesel combustion engine a large amount of energy is removed from the exhaust gases in order to lower both the gas temperature and oxygen content. Thereby, the EGR system allows significant reduction of nitrogen oxides (NO_x) in the combustion process but it also produces waste energy. Part of this waste energy can be converted to useful electric energy using thermoelectric modules. Recently, several experimental studies on thermoelectric systems for heat recuperation in automotive applications has been studied and presented in the literature [13–20]. Although the theoretical basis for modeling fluid dynamics and thermoelectricity is well established, first principle simulations cannot be conducted for the foreseeable future. This is due to the complexity of the problem and inherent problem with separation of scales, e.g., electrical and thermal contact resistances on the smallest scales, geometric shape of connectors and semiconducting material on the intermediate scale (millimeter), up to the size of heat exchanger (meter scale). Only for very small systems it is possible to do detailed multiphysics simulation where fluid dynamics is combined with a model of the thermoelectric generator (TEG) based on first principle simulations, i.e., solving the thermoelectric constitutive equations inside the TE modules [21,22].

On the contrary, systems studied in the literature often relate to energy recovery applications that may consist of tens to hundreds of thousands interior parts with scales separated several orders of magnitudes preventing first principle simulations, even on the largest computer clusters. First principles thermoelectric calculations require full resolution of the temperature field, solution of the electric charge continuity equation, and that the thermoelectric constitutive equations are solved for ten to hundreds of thousands interior parts. Similarly, first principles computational fluid dynamics (CFD) simulations require solution of the Navier–Stokes equation, but for turbulent flow simulation this is too costly at high Reynolds numbers. Instead turbulence models are introduced, and this filtering operation means that the Reynolds-averaged Navier–Stokes equation is solved instead. Therefore, the models of the thermoelectric performance in heat recovery systems has been more or less simplified, thereby to certain extent sacrificing physics. Weng and Huang developed models for studying heat exchangers attached with thermoelectric modules for waste heat recovery [23]. In their analysis, only open-circuit systems were simulated. Thereby the effect of electric current on the heat transport e.g., Joule, Peltier and Thomson effects was excluded from the analysis [23]. Martinez and co-workers used CFD simulations to design heat exchangers and to calculate the pressure drop in TEG heat exchanger [24]. The authors have analyzed heat transfer resistances in the heat exchangers and in thermoelectric modules to identify rate limiting steps that affect TEG efficiency [7]. Tang et al. developed a model to simulate the flow field and temperature distribution inside a heat exchanger for an automotive thermoelectric generator [25]. For this purpose, CFD simulations was used, but the thermoelectric power generation was not integrated in the analysis. Consequently, the electrical current and its effect on the heat flow and temperature distribution could not be determined. Su et al. reduced the complexity and analyzed system efficiency with respect to heat transfer without coupling the CFD model to the thermoelectric power generation [26]. Wang et al. designed and optimized a heat exchanger for TEG application using CFD analysis also without describing effects of Joule, Peltier and Thomson effects [27]. There are also examples of studies where the complexity is reduced by neglecting the fluid dynamics simulation thereby simplifying the analysis significantly. In these studies the computational effort is rather directed towards predicting the thermoelectric power generation as in the work by Deng [28]. Similar simplification is found in the work on understanding the transients and startup of TEGs by Yu [29], who studied the heat flow caused by thermoelectric effects including Peltier, Joule and Thomson heat, but without doing simultaneous CFD analysis. The need for a physical and at the same time computational efficient model is apparent, as it will facilitate design and detailed analysis of energy recover systems in the future.

Development of a generic multiphysics model that allows the coupling between fluid dynamics simulations and thermoelectricity to be predicted is therefore addressed in the current study. For this purpose, the authors extend a modeling framework proposed for energy flow in large networks of thermoelectric modules [30] and couple these models with state-of the-art CFD simulations for turbulent flow heat-transfer. The objective of developing the present multiphysics model is therefor to allow state of the art fluid dynamics modeling to be done in conjunction with thermoelectric modeling by accounting for the two-way coupling of energy in the governing equations. Additionally, the objective of the present work is to validate the new model with experiments under conditions relevant to energy recovery applications and to evaluate its accuracy.

2. Experimental Methodology

The TEG heat exchanger developed within this work is primarily intended for validation of the new simulation methodology. It consists mainly of two parts as shown in Figure 1. The lower part includes the gas channel where the energy in the hot gas is transported to the upper part—which also includes the thermoelectric modules by use of heat pipes.

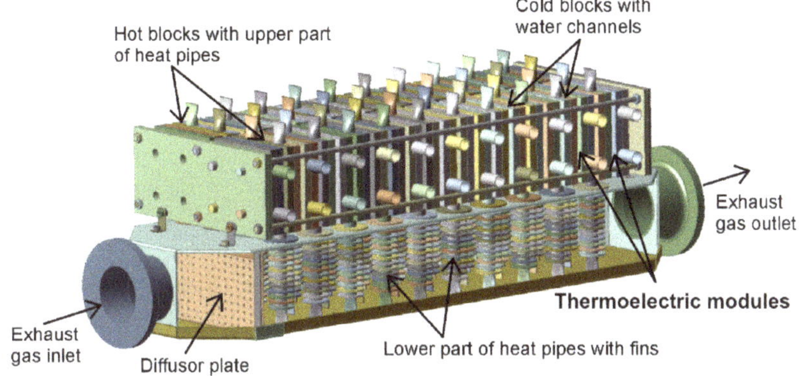

Figure 1. Design of the thermoelectric generator (TEG) heat exchanger.

In order to provide a minimum of heat transfer resistance, the heat pipes were made of copper which has excellent thermal conductivity compared to other metals. A corresponding photo of the hardware is shown in Figure 2. For the purpose of achieving a useful voltage level, the 40 modules were connected electrically in five groups in series, each containing 8 modules and the groups were connected together in parallel. In this work, 40 commercial Bi2Te3 modules TEPH1-12680-0.15 from Thermonamics Electronics Co., Ltd. are installed in the TEG heat exchanger. Information about the module efficiency, structure and the electric and thermal properties of the material as function of temperature is provided in the paper by Högblom and Andersson [31] The diffusor plate was designed with increasing size of the holes towards the sides of the plate to partly counteract flow separation and provide uniform heat transfer to the pipes. For the same reason, the heat pipes were design with an increasing number of fins downstream the heat exchanger to provide a uniform heat transfer in the stream wise direction since the gas enthalpy decrease.

Figure 2. Photo of the TEG heat exchanger hardware.

The measurements were done in the engine bench laboratory using the exhaust gas from a one cylinder, 2.1-liter diesel engine. Besides engine speed and torque the exhaust gas temperature and flow rate are affected by fuel injection parameters. Three different load points were measured where engine speed and torque were varied resulting in different gas temperatures, mass flow rates and thereby different electrical power generation. During all the experiments the gas temperature, mass flow rate and pressure drop was measured together with the total voltage and current from the TEG unit after sufficiently long time, when steady state conditions was reached. In the following text and figures the three load points are referred to as the low, moderate and high load cases, due to the difference in electrical power generation as shown in the last column in Table 1. The thermoelectric system was connected to an electronic load (Amrel, PLA 1.5K-60-600) where the current and voltage were measured at the different engine load point with an accuracy of 0.05%. Temperatures measurements in thermoelectric heat exchanger were accomplished with a temperature module (National Instruments, CT-120), having a measurements accuracy of 0.05%. Load resistance was chosen to maximize the electrical power output, i.e., load resistance equals the total resistance of the system.

Table 1. Engine bench load points used for model validation.

Load Point	Mass Flow (g/s)	Gas Temp., Inlet (C)	Electric Power (W)
Low	78.56	423.1	137.1
Moderate	79.99	459.2	174.7
High	88.01	489.4	212.6

3. Simulation Methodology

The requirements for the multiphysics model developed in this work include models to predicting the rate of turbulent heat transfer and models to predict the thermoelectric power generation and at the same time allow full coupling between the different phenomena. In Figure 1 the front wall of the heat exchanger is removed revealing the interior parts. As shown here the heat transfer from the gas is enhanced by using heat pipes covered by fins. Since the performance of a TEG system is strongly dependent on the temperature difference over the thermoelectric material all thermal resistance from the heat source to the heat sink must be minimized. This can be accomplished by heat pipes, which are sealed tubes containing a gas-liquid mixture that allow large amounts of energy to be transferred with minimal temperature drop by the process of evaporation and condensation. For this purpose, heat pipes have been used together with TEGs applications both at the hot and the cold side [1,20,32–34]. Kim et al. and Jang et al. built and analyzed TEG systems containing heat pipes for automotive applications [20,33]. The present heat exchanger contains a total of 22 cold blocks, 20 hot blocks and 40 thermoelectric modules, thereby 2 modules are installed between each cold and hot block. Each hot block is connected to two heat pipes and two thermoelectric modules as shown in Figure 1.

Accurate description of the heat transfer process from the hot gas to the fins in the heat exchanger requires a turbulence model that allows high resolution close to the fins where the thermal boundary layers develops. This means that turbulence models such as the k-ε model with wall functions are not appropriate to use due to the fact that the flow is confined between narrow fins influenced by viscous forces which prevents the use of wall functions as they can only be used when the first grid point is located in the inertial sublayer, i.e., $y^+ \geq 30$ [35].

For the propose of resolving the turbulent boundary layers and the heat flux correctly the SST k-ω is used. This model is usable all the way through the buffer and viscous sublayers without introducing any turbulence damping function, which is required if a k-ε models should be used in the near wall region [36]. Since the SST k-ω model allows calculation throughout the near wall region including the viscous sublayer, it can resolve regions of different wall stresses thereby allowing the thin thermal boundary layers to be resolved correctly in all parts of the heat exchanger, as long the discretization is dense enough in the wall region. In these simulations the fluid can safely be treated as incompressible gas, using the ideal gas law since the Mach number is approximately 0.1, and the effects due to intramolecular interactions between the molecules are negligible at all temperature and pressure conditions inside the system. The time-filtered continuity equation reads

$$\frac{\partial \langle U_i \rangle}{\partial x_i} = 0 \tag{1}$$

and the conservation equation for momentum transport, the Reynolds averaged Navier–Stokes equation, is given by

$$\frac{\partial \langle U_i \rangle}{\partial t} + \langle U_j \rangle \frac{\partial \langle U_i \rangle}{\partial x_j} = -\frac{1}{\rho}\frac{\partial}{\partial x_j}\left\{ \langle P \rangle \delta_{ij} + \mu \left(\frac{\partial \langle U_i \rangle}{\partial x_j} + \frac{\partial \langle U_j \rangle}{\partial x_i} \right) - \rho \langle u_i u_j \rangle \right\}. \tag{2}$$

Here the Reynolds stress term, $\rho \langle u_i u_j \rangle$, is closed by the Boussinesq approximation [35], using information about the local turbulent properties, i.e., local length and velocity scales of turbulence, determined by the SST k-ω turbulence model. The SST k-ω model is a two equation turbulence model which allows the advantages of the classical k-ω model (good performance in the near wall region) and k-ε model (for the free stream) to be retained in one and the same model. This makes the SST k-ω model accurate and computationally efficient [36]. In the SST k-ω model, the transport equation for turbulent kinetic energy, k, is given by

$$\frac{\partial k}{\partial t} + \langle U_i \rangle \frac{\partial k}{\partial x_i} = \frac{\partial}{\partial x_i}\left(\Gamma_k \frac{\partial k}{\partial x_i} \right) + P_k - D_k \tag{3}$$

and the transport equation for the specific rate of energy dissipation, ω, is given by

$$\frac{\partial \omega}{\partial t} + \langle U_i \rangle \frac{\partial \omega}{\partial x_i} = \frac{\partial}{\partial x_j}\left(\Gamma_\omega \frac{\partial \omega}{\partial x_i} \right) + P_\omega - D_\omega + C_\omega \tag{4}$$

Here Γ is the effective diffusivity, P is the production, D is the dissipation of the corresponding variables and C_ω is the cross-diffusion term. There are some differences between the classical k-ω model and the SST k-ω, for example in calculation of the production term [36]. The terms describing the production and dissipation of ω as well as the cross-diffusion term also contains a blending function that allows a smooth transition between the k-ω model performance in the near wall regions and the k-ε model performance in the bulk flow. This blending function is determined as a function of the distance to the nearest surface as described by Menter [36]. The energy transport is modeled by

$$\frac{\partial (\rho c_p T)}{\partial t} + \langle U_i \rangle \frac{\partial (\rho c_p T)}{\partial x_i} = \frac{\partial}{\partial x_i}\left(\frac{\nu}{\mathrm{Pr}} + \frac{\nu_T}{\mathrm{Pr}_T}\frac{\partial T}{\partial x_i} \right) + S_{rad} \tag{5}$$

where $\mathrm{Pr} = \nu/\alpha$ and the heat diffusivity $\alpha = \lambda/(\rho c_p)$ and Pr_T is the turbulent Prandtl number. Heat transfer at the boundaries are negligible due to insulation and imposed using zero heat flux boundary conditions in the model, except for boundaries in contact with the thermoelectric modules. Radiative heat transfer between surfaces of different temperature inside the heat exchanger is include in the model. In this work, the discrete ordinates (DO) radiation method is used to solve the radiative transfer equation (RTE) numerically [37] and a transport equation is solved for the radiation intensity in the spatial coordinates,

$$\nabla \cdot \left(I(\vec{r}, \vec{s})\vec{s} \right) + aI(\vec{r}, \vec{s}) = an^2 \frac{\sigma T^4}{\pi} \tag{6}$$

Here the radiation absorption coefficient for the gas is calculated used weighted-sum-of-gray-gases (WSGG) model [37].

Besides Fourier heat conduction to the thermoelectric modules, the effect of Peltier, Thomson and Joule heating in the semiconducting material needs to be accounted for in the CFD model. The complexity of this modelling is twofold. Firstly, it is largely affected by the current in the semiconducting material which implies that it is also affected by the electrical circuit i.e., electric connections between the different thermoelectric modules in the system. Secondly the Peltier, Thomson and Joule heating effects the local temperature and the heat flux in the system causing effects on heat transfer over the entire system that should be handled by the CFD model, expect in the case for open circuit but that is of no interest for heat recovery systems. Consequently, the problem requires two way coupling between the CFD and thermoelectric models allowing the effect of current and heat transfer over the entire heat exchanger to be included in the analysis. The generic formulation of the thermoelectric models consists of three sub models. This includes electrical and thermal models for the individual modules and one model for the circuit of the connected system. Derivation, discussion and validation of these models, over a wide range of operating conditions, even when some of the modules in the network work with reversed current, is found in [30]. The electrical model contains modification to the electrical model proposed by Montecuccu et al. [38], which for individual modules in the network is given by

$$U = \left(\beta_{s1} + \beta_{s2}T_{avg}\right)\Delta T - I\left(\beta_{R1} + \beta_{R2}T_{avg} + \beta_{R3}T_{avg}^2\right) \tag{7}$$

In Equation (7) the Seebeck voltage and the internal resistivity is accounted for, the latter is affected by both the electrical contact resistances in material junctions and the materials resistivity. In Equation (7), U and I is the voltage over and current through the module, respectively, T_{avg} and ΔT are the average temperature in the semiconducting material and the temperature difference between the hot and the cold blocks adjacent to the module. Here β coefficients are regression parameters determined for the module using the method described in [30]. Furthermore, the heat flow on the cold side of the for individual modules are given by

$$Q_c = \left(\beta_{F1} + \beta_{F2}T_{avg}\right)\Delta T + I\left(\beta_{PT1}T_c + \beta_{PT2}T_{avg}\Delta T\right) + \frac{I^2\left(\beta_{R1} + \beta_{R2}T_{avg} + \beta_{R3}T_{avg}^2\right)}{2} \tag{8}$$

The total heat flow in Equation (8), is described by Fourier heat conduction along with the Peltier and Thomson effect and the Joule heating. In contrast the heat flow on the hot side of the modules also contains an additional term for the heat that is converted to electrical energy

$$Q_h = \left(\beta_{F1} + \beta_{F2}T_{avg}\right)\Delta T + I\left(\beta_{PT1}T_c + \beta_{PT2}T_{avg}\Delta T\right) + I\left(\beta_{s1} + \beta_{s2}T_{avg}\right)\Delta T - \frac{I^2\left(\beta_{R1} + \beta_{R2}T_{avg} + \beta_{R3}T_{avg}^2\right)}{2}. \tag{9}$$

As a result, the difference between the heat flow to the module surface on the hot side and from the module surface on the cold side corresponds to the net electric power generated by the thermoelectric modules. A general circuit model for the connected system based on Ohms law and Kirchhoff's current law is presented in [30]. In this case the thermoelectric generator used for validation has 40 modules (8×5) and the equations becomes

$$U_{tot} = \sum_{j=1}^{8} U_{ji} \qquad \forall\, i \in \{1,\ldots,5\} \tag{10}$$

and

$$I_{load} = \frac{U_{tot}}{R_{load}} = \sum_{i=1}^{5} I_i \tag{11}$$

where index, i, represent the 5 different serial connected groups and index, j, represent the each of the 8 modules within the groups. By solving of Equations (1)–(11) a two-way coupled energy transport is made possible by using the thermoelectric models as subgrid models in CFD analysis and by assigning heat flow boundary conditions on the thermoelectric module surfaces in the CFD model.

A block diagram describing the coupling and how the thermoelectric models are implemented in the CFD analysis is shown in Figure 3. As shown here the temperature distribution over the different modules effect the electrical performance. This is in turn affected by the module current which is predicted by the circuit model. Knowledge of the current distribution allows Peltier, Thomson and Joule effects to be quantified when passing back information on the heat flow to the CFD solver. Hence, the circuit model allows local variations in driving force to be sensed throughout the entire system of thermoelectric modules. The flow of hot gases and the turbulent heat transfer were modeled using commercial CFD solver—Ansys Fluent 19—and the three thermoelectric models are implemented in the CFD analysis as subgrid programs written in C. Every iteration, the temperature inside the hot and cold blocks adjacent to all the modules are calculated by the program and the Seebeck voltages and internal resistances are calculated in the electrical model. Together with the load resistance and the electrical connections the currents through all modules are determined in an iterative manner in the circuit model and the electrical model pass on information allowing the thermal effects to be calculated by the thermal model, Equations (7) and (8). This information is passed back to the CFD solver to be used as heat flux boundary condition in the next iteration as shown in Figure 3. Thereby the effect of current on the heat flow in the CFD analysis is accounted for throughout the entire heat exchanger, providing a full two-way coupling through energy conservation.

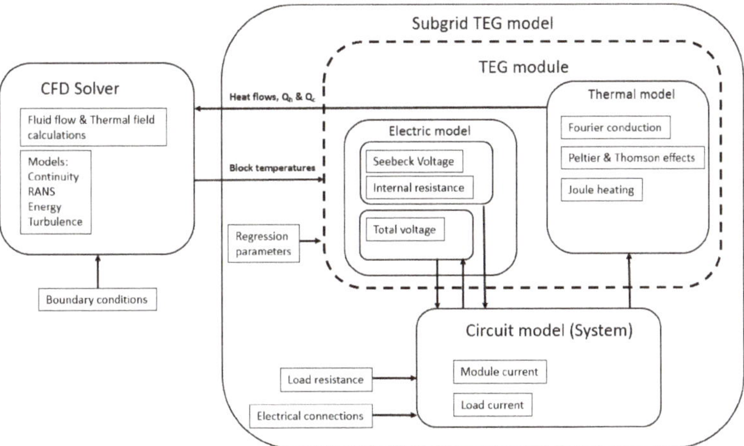

Figure 3. Schematic of program structure and calculation procedure.

In other words, by allowing the flow field and temperature fields to be solved simultaneously with the thermoelectric power generation—and by accounting for the Peltier, Thomson and Joule effects—the multiphysics model presented here provides an efficient and correct modeling approach to handle an arbitrary number of thermoelectric modules in CFD analysis in a physical and robust way.

3.1. Mesh Structure, Numeric Methods and Sensitivity Analysis

High quality heat-transfer simulations require good mesh quality and suitable numeric schemes to control numeric errors. For this purpose, prism layers were introduced in all boundary layers and second–order discretization schemes were used to minimize numeric errors. Since the geometry of the computational domain is complex, the geometry was divided into different zones allowing generation of a swept mesh and high-quality hexahedral cells as depicted in Figure 4.

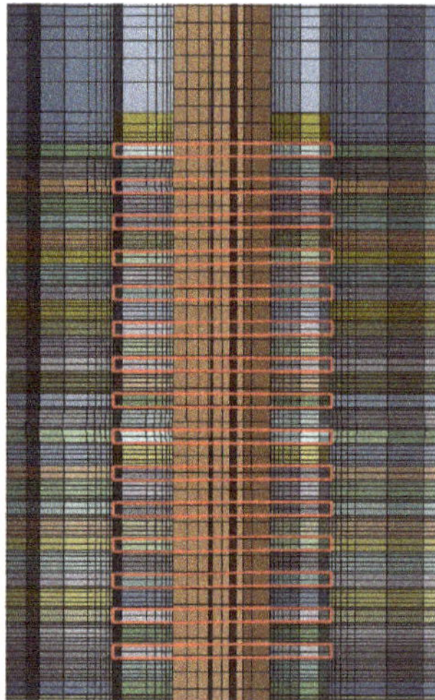

Figure 4. Cross section of the computational mesh through one of the heat pipes.

Mesh independence was confirmed by evaluating the solution variables using models with different mesh resolution. The mesh adaption was done by refining mesh zones with large gradients of the solution variables. Both turbulent kinetic energy and the thermal field have the largest gradients in the near wall regions which is consistent with the main heat transfer resistance being in the gas film close to the surfaces. Evaluation of mesh independence showed that mesh independent results were obtained with a model containing 4.2 million cells. This mesh was therefore used for all the simulations in this study. Simulations were done on a computer cluster using up to 128 cores.

Furthermore, a sensitivity analysis with respect to the turbulence model was done. The SST k-ω turbulence model is judged the best model in this case, due to the reasons discussed earlier, but turbulence simulations using dedicated low Reynolds k-ε models can also give good results. Similar to the SST k-ω this model also allows integration throughout the near wall region. This is achieved by using turbulence dampening functions that are active in the buffer and viscous sublayers. In this case the Launder-Sharma low Reynolds k-ε model was used [39]. For both SST k-ω and Launder-Sharma models the y^+ -value were sufficiently low on all interior walls. It was found the difference between the turbulence models is not critical since the difference was very small, lower than 1% for the predicted pressure drop and heat transfer rates. Finally, a sensitivity analysis was made with respect to the turbulence boundary conditions as discussed in the section below.

3.2. Boundary Conditions and Material Data

The boundary condition given in Table 1, are specified according to the mass flow rates obtained at the different engine load cases. Turbulence boundary conditions at the inlet were given as turbulent intensity and the turbulent length scale. These properties were determined based on best practice guidelines, i.e., turbulence intensity was estimated as $0.16Re^{-1/8}$ and the length scale as 7% of the hydraulic diameter [35]. A sensitivity analysis of the turbulence boundary conditions was done by changing these values by 10%. It was found the boundary conditions had no significant effect on the final results (<0.1%). These results can be understood better by analyzing the rate of turbulence dissipation and turbulence production. The analysis revealed turbulence in the inlet does not survive long enough to have any impact after the diffusor plate, shown in Figure 1. At the diffusor plate new turbulence is continuously generated which dominate the rate of momentum and heat transfer downstream in the system, thereby making the simulations less sensitive to the turbulent boundary conditions applied at the inlet. At the exhaust gas outlet, a constant pressure boundary condition was implemented. The walls of the well-insulated heat exchanger were treated adiabatically using a zero heat flux boundary condition.

The temperature dependency of the specific heat, viscosity and thermal conductivity of the gas were given by NIST data. Analysis of the heat transfer rate through the heat pipes shows that this is not a limiting factor for the overall heat transfer rate from the exhaust gas bulk to the surface of the hot block, instead the main heat transfer resistance lies in the gas film. This allows, the heat transfer in the heat pipes to be simulated using an effective heat conductivity. The same modeling strategy for heat pipes has been presented in the literature and shown to give accurate results [40]. This was also validated by measurements of heat flow in heat pipes for the temperatures relevant for the system, which shows all conditions are far from being controlled by the internal heat transfer rate inside the heat pipes. To finally close the model, there are thermal contact resistances present between the heat pipes and the hot aluminum blocks and between the heat pipes and the fins, which are accounted for in these simulations. The contact resistances specified in these simulations were determined in a previous study for similar material contacts [31], which is consistent with values reported in the literature [41].

4. Results and Discussion

The results presented here are based on mesh independent simulations using the SST k-ω model for the three engine load points introduced earlier. Validation of the simulation model includes both the fluid dynamics and the thermoelectric performance obtained by solving Equations (1)–(11). The simulations were run on a computer cluster until convergence of all solution variables and by ensuring mass and energy conservation at the end of the iterations. Figure 5 shows the temperature field inside the heat exchanger in a horizontal plane in the middle section overlaid with streamlines colored by the residence time. The heat transfer to the heat pipes and the thermoelectric modules causes the temperature to decrease in the exhaust gas flow, mainly in the center whereas the flow along the outermost streamlines are less affected. Due to insulation at the walls the gas temperature in the wall region only decrease as a result of turbulent mixing with the bulk fluid that has lower temperature. When analyzing the temperature distribution at the outlet in Figure 5, it becomes clear that the gas is not fully mixed, i.e., hot gas flows along the pipe walls and cold gas in the center. Therefore, temperature measurements at this location can be sensitive to the location of the sensor. Contrary, pressure measurements would not be sensitive since radial variations in pressure is not significant. From an energy balance point of view approximately one third of the enthalpy is transferred to the modules. Out of this energy a fraction is converted to electric energy in the thermoelectric modules.

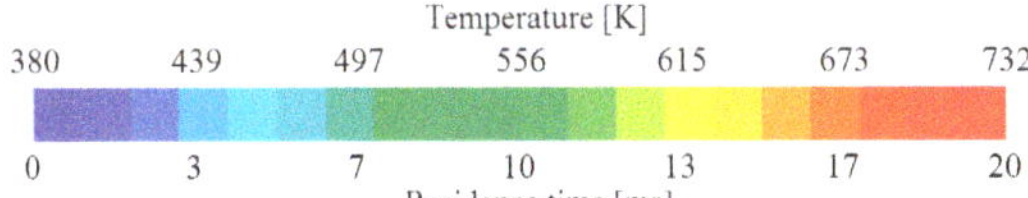

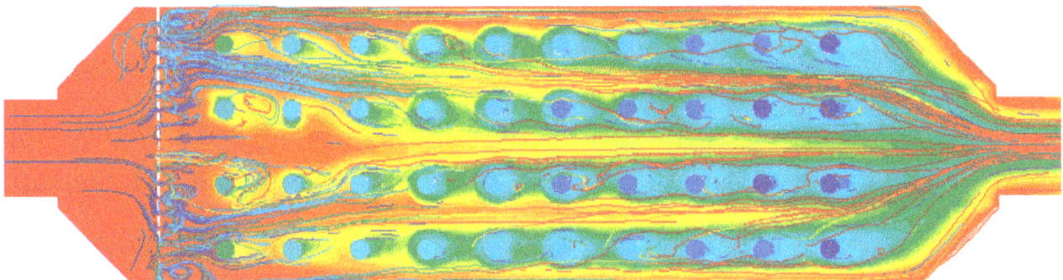

Figure 5. Streamlines and temperature distribution in a horizontal plane at moderate engine load.

The temperature distribution also results in differences in local heat transfer rates and thereby a distribution of surface temperatures between the hot aluminum blocks. Figure 6 displays the temperature distribution on the heat pipe surfaces and the temperature of the hot and cold aluminum blocks, respectively. Here the effect of the inlet diffuser plate on heat transfer is clearly visible. The temperature difference between the first and second row of heat pipes is much larger compared to the following rows. This is explained by that the turbulent jets formed by the diffuser plate does not penetrate long enough to have significant effect beyond the second row of heat pipes. It is also noticeable how the diffuser plate, which is designed with smaller holes in the middle part and larger holes at the sides actually provides higher heat transfer rate to the outermost heat pipes, which are 40 °C warmer compared to the two heat pipes in the center. However, this difference is not fully sensed by the thermoelectric modules, since the metal blocks to which modules are attached, have high thermal diffusivity that allows lateral heat flow which minimize the temperature distribution over the module surfaces. To counteract the problem with low heat transfer rate in gases and simultaneously provide more homogenous thermal load on the thermoelectric modules, the heat pipes are designed with gradually increasing number of fins. As shown in Figure 6, the first three rows of heat pipes at the inlet have 13 fins each, the next three rows have 16 fins and the last four rows are designed with 20 fins each to compensate for the temperature drop in the stream wise direction. The simulations reveal that the present design does not entirely compensate for this and the model provide detailed enough information to allow optimization of the design.

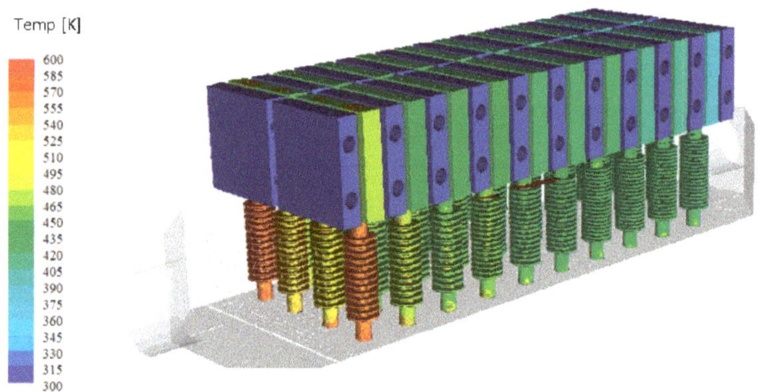

Figure 6. Temperature distribution on heat pipe surfaces and aluminum blocks, high-engine-load case.

Validation of pressure drop and temperature drop for the three engine load points at steady state engine bench conditions is summarized in Figure 7. Overall, the simulation results agree very well at the three engine loads points. Considering the calculated pressure drop over the thermoelectric heat exchanger the error is only 2.3%, 0.4% and 0.4% for the different engine loads, as shown in Figure 7a. Figure 7b shows the temperature drop, defined as the difference between the inlet and outlet exhaust gas temperature. Predictions of the temperature drop is considered good, but slightly larger, 6.0%, 4.6% and 4.8% in the three cases. Considering the complexity of the geometry and that the flow contains bot turbulence and regions affected by viscous forces including turbulent jets at the diffusor plate this is considered as very good results. As shown Figure 7b, there is small and systematic deviation that can be due to that a thermocouple installed in the gas flow may register a slightly too high gas temperature since the sensor is influenced by radiation from the hot surrounding wall. The sensor is installed in the center of the pipe where the temperature is lower compared to the surrounding wall. If a slightly too high temperature is measured by the sensor at the outlet it follows that the temperature drop over the system gets too small. Consequently, measurement points, in Figure 7b, will be lower than they should. Calculation accounting for both convective heat transfer and radiative heat transfer to the thermocouple indicates that half the difference seen in Figure 7b can be explained by this effect. In other words, half the difference shown in Figure 7b can be explained by this phenomenon. The accuracy of the predications is therefore closer to 3% rather than 6% as seen at first. The use of radiation shielded thermocouples, can solve this problem or an approach using the two-thermocouple method can be used to minimize the difference [42].

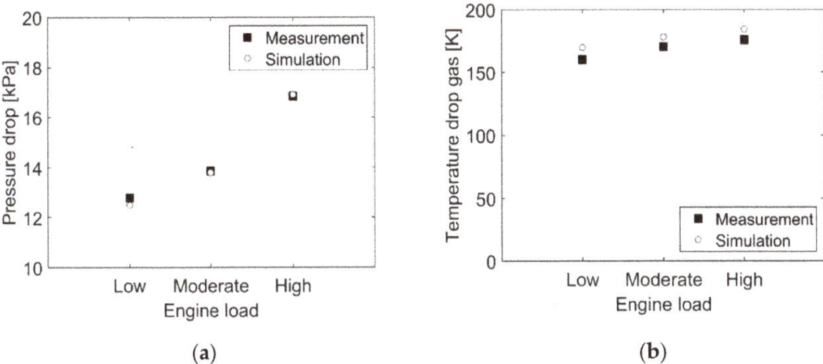

Figure 7. Comparison between experiments and simulation results. (a) Pressure drop; (b) temperature drop.

In order to evaluate the accuracy of the predicted electric power generation, the current and voltage are compared with measurements at the three load points. Figure 8a,b summarizes the results for the thermoelectric power generation and shows that the predicted current and voltage agrees very well with the measurements for all engine loads. The error is less than 1.5% for all experimental conditions and no systematic error exists.

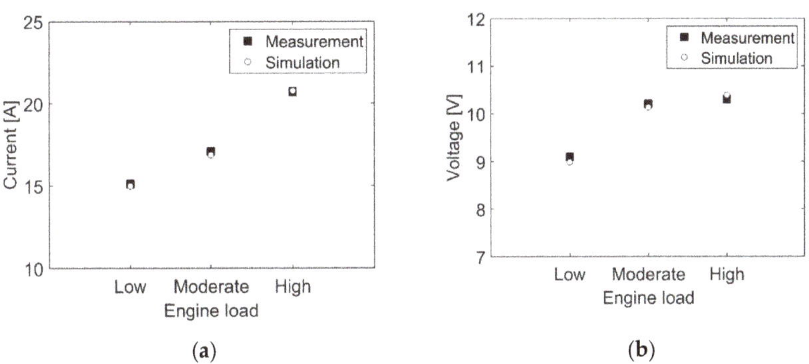

Figure 8. Comparison between experiments and simulation results. (a) Current; (b) voltage.

Overall, the agreement between the simulation results and measurements is good. This leads to the conclusion that the multiphysics CFD model accounts correctly for both fluid dynamics, including pressure drop and heat transfer and the thermoelectric power generation. At the same time, it means that multiphysics model allows the importance of two way coupling to be analyzed and understood particularly since the thermoelectric modules operate at different temperature while being electrically connected. The temperature distribution obtained in these simulations were therefore used to analyze a case where all modules are allowed to operate individually at the current maximizing their power output to assess the impact on the electrical power output of the TEG system. This gives an estimate of the highest power output achievable at these thermal conditions and provides a direct measure of the magnitude of the non-ideal effect caused by the electrical connection and consequently also the error that is introduced if the two way coupling is neglected in evaluation of the TEG performance. Considering the three engine load cases such simulation approach leads to overprediction of the power output corresponding to 5.4%, 5.2% and 4.6%, respectively. These are significant deviations compared to the accuracy achieved with the proposed model which gives less than 1.5% difference.

The error introduced by simplistic analysis could obviously be much larger when the temperature distribution is more significant which is expected from systems working at higher heat recovery levels. An improved heat exchanger design that allows more energy to be transfer from the gas results in a larger temperature difference among the thermoelectric modules and thereby also larger difference in heat flow and Seebeck voltage in the different modules. Consequently, the non-ideal effects would increase substantially. In order to understand this effect better at higher heat recovery level the system was simulated with increased heat transfer. The results are summarized below in Table 2.

Table 2. Electrical power output as function of heat recovery level.

Transferred Energy (%)	Real Power (W)	Optimal Power (W)	Difference (%)
34.0%	136.2	142.5	4.6
56.0%	335.8	379.6	13.0

As shown in Table 2, the difference between the optimal (fictious) power and the real power generation increase rapidly in a system of connected modules with increasing heat recovery level. This is due to the electrical connection between the modules inside the thermoelectric generator and increasing temperature differences in the system. A simplistic modeling approach would therefor lead to unnecessary large simulation error which highlights the importance of using a coupled approach in simulating the heat transfer and thermoelectric power generation. Considering that the methodology presented here is computational efficient and generic it may allow accurate prediction of power generation in other systems and concepts by accounting for effect of the current on the heat flow throughout the systems.

5. Conclusions

A multiphysics simulation methodology was developed and validated that permits the extension of computational fluid dynamics (CFD) simulations to account for electric power generation and its effect on the energy transport, the Seebeck voltage, the electrical currents in thermoelectric systems. The energy transport through Fourier, Peltier, Thomson and Joule mechanisms as function of temperature and electrical current—and the electrical connection between thermoelectric modules—was integrated in the CFD analysis by subgrid models. Thereby the effect of current on the heat flow in the CFD analysis is accounted for throughout the entire heat exchanger, providing a full two-way coupling through energy conservation. The accuracy was evaluated for a thermoelectric generator containing 40 connected thermoelectric modules in engine bench laboratory and showed that the fluid dynamics and the thermoelectric performance were predicted very well. Comparison with simulation results showed the predicted pressure drop has an error of 2.3%, while the temperature drop over system has an error of 3.0%, at all experimental conditions studied, when accounting for the radiative effects on the thermocouple. The thermoelectric power generation was predicted with high accuracy and the error in voltage and current was less than 1.5% at the exhaust gas flow rates and temperatures studied. Simulation results confirm computational efficiency and stable simulations with low increase in computational time compared to standard CFD heat-transfer simulations. Furthermore, the importance of two way coupling of energy transport was analyzed. Evidence was found that even at the lowest heat transfer rate studied it is required to use a full two way coupling in the energy transport to accurately predict the electric power generation.

Author Contributions: Methodology O.H., R.A.; simulations O.H.; validation, O.H., R.A; writing, O.H., R.A.; project administration R.A.; funding acquisition, R.A. All authors have read and agreed to the published version of the manuscript.

Funding: This work was supported by the Swedish Foundation for Strategic Environmental Research.

Acknowledgments: The financial support from the Swedish Foundation for Strategic Environmental Research is gratefully acknowledged. The authors would like to thank Termo-Gen AB building the prototype and Volvo Technology AB for doing the engine bench measurements. The multiphysics simulations were done using the high-performance computer cluster at Chalmers University of Technology, supported by the Swedish National Infrastructure for Computing, SNIC.

Conflicts of Interest: The authors declare no conflict of interest.

Nomenclature

α	Seebeck coefficient, V K^{-1}
c_p	Heat capacity, J kg^{-1} K^{-1}
ε	Dissipation rate of turbulent kinetic energy, m^2 s^{-3}
I	Current, A
I	Radiation intensity, J m^{-2} s^{-1}
k	Turbulent kinetic energy, m^2 s^{-2}
λ	Heat conductivity, W m^{-1} K^{-1}
μ	Dynamic viscosity, Pa s
ν	Kinematic viscosity, m^2 s^{-1}
Π	Peltier coefficient, V
Ψ	Combined Peltier and Thomson coefficients, V
P	Pressure, Pa
Q	Source term, W m^{-3}
Q	Heat flow, W
ρ	Density, kg m^{-3}
R	Resistance, Ω
T	Temperature, K
τ	Thomson coefficient, V K^{-1}
U	Voltage, V
U_i	Velocity, m s^{-1}
ω	Specific dissipation rate of turbulent kinetic energy, s^{-1}
y^+	Dimensionless wall distance, -

Subscripts

avg	Average between hot and cold side
c	Cold side
h	Hot side
int	Internal
L	Load
Si	Seebeck where $i = 1, 2$
Ri	Internal resistance
Fi	Fourier conduction
PTi	Peltier & Thomson
T	Turbulent

References

1. He, W.; Su, Y.; Riffat, S.; Hou, J.; Ji, J. Parametrical analysis of the design and performance of a solar heat pipe thermoelectric generator unit. *Appl. Energy* **2011**, *88*, 5083–5089. [CrossRef]
2. Ji, J.; Lu, J.-P.; Chow, T.-T.; He, W.; Pei, G. A sensitivity study of a hybrid photovoltaic/thermal water-heating system with natural circulation. *Appl. Energy* **2007**, *84*, 222–237. [CrossRef]
3. Zhang, M.; Miao, L.; Kang, Y.P.; Tanemura, S.; Fisher, C.A.J.; Xu, G.; Li, C.X.; Fan, G.Z. Efficient, low-cost solar thermoelectric cogenerators comprising evacuated tubular solar collectors and thermoelectric modules. *Appl. Energy* **2013**, *109*, 51–59. [CrossRef]

4. Suter, C.; Jovanovic, Z.R.; Steinfeld, A. 1 kWe thermoelectric stack for geothermal power generation—Modeling and geometrical optimization. *Appl. Energy* **2012**, *99*, 379–385. [CrossRef]

5. Whalen, S.A.; Dykhuizen, R.C. Thermoelectric energy harvesting from diurnal heat flow in the upper soil layer. *Energy Convers. Manag.* **2012**, *64*, 397–402. [CrossRef]

6. Espinosa, N.; Lazard, M.; Aixala, L.; Scherrer, H. Modeling a Thermoelectric Generator Applied to Diesel Automotive Heat Recovery. *J. Electron. Mater.* **2010**, *39*, 1446–1455. [CrossRef]

7. Högblom, O.; Andersson, R. CFD Modeling of Thermoelectric Generators in Automotive EGR-coolers American Institute of Physics. *Conf. Ser.* **2011**, *1449*, 497–500.

8. Wang, Y.; Dai, C.; Wang, S. Theoretical analysis of a thermoelectric generator using exhaust gas of vehicles as heat source. *Appl. Energy* **2013**, *112*, 1171–1180. [CrossRef]

9. Li, B.; Huang, K.; Yan, Y.; Li, Y.; Twaha, S.; Zhu, J. Heat transfer enhancement of a modularised thermoelectric power generator for passenger vehicles. *Appl. Energy* **2017**, *205*, 868–879. [CrossRef]

10. Nonthakarn, P.; Ekpanyapong, M.; Nontakaew, U.; Bohez, E.L.J. Design and Optimization of an Integrated Turbo-Generator and Thermoelectric Generator for Vehicle Exhaust Electrical Energy Recovery. *Energies* **2019**, *12*, 3134. [CrossRef]

11. El-Genk, M.S.; Saber, H.H. Performance analysis of cascaded thermoelectric converters for advanced radioisotope power systems. *Energy Convers. Manag.* **2005**, *46*, 1083–1105. [CrossRef]

12. Cho, Y.H.; Park, J.; Chang, N.; Kim, J. Comparison of Cooling Methods for a Thermoelectric Generator with Forced Convection. *Energies* **2020**, *13*, 3185. [CrossRef]

13. In, B.D.; Lee, K.H. A study of a thermoelectric generator applied to a diesel engine. *Proc. Inst. Mech. Eng. Part. D J. Automob. Eng.* **2015**, *230*, 133–143. [CrossRef]

14. Liu, X.; Deng, Y.; Li, Z.; Su, C. Performance analysis of a waste heat recovery thermoelectric generation system for automotive application. *Energy Convers. Manag.* **2015**, *90*, 121–127. [CrossRef]

15. Liu, X.; Li, C.; Deng, Y.; Su, C. An energy-harvesting system using thermoelectric power generation for automotive application. *Int. J. Electr. Power Energy Syst.* **2015**, *67*, 510–516. [CrossRef]

16. Tang, Z.B.; Deng, Y.D.; Su, C.Q.; Shuai, W.W.; Xie, C.J. A research on thermoelectric generator's electrical performance under temperature mismatch conditions for automotive waste heat recovery system. *Case Stud. Therm. Eng.* **2015**, *5*, 143–150. [CrossRef]

17. Thacher, E.F.; Helenbrook, B.T.; Karri, M.A.; Richter, C.J. Testing of an automobile exhaust thermoelectric generator in a light truck. *Proc. Inst. Mech. Eng. Part. D J. Automob. Eng.* **2007**, *221*, 95–107. [CrossRef]

18. Hsu, C.-T.; Huang, G.-Y.; Chu, H.-S.; Yu, B.; Yao, D.-J. Experiments and simulations on low-temperature waste heat harvesting system by thermoelectric power generators. *Appl. Energy* **2011**, *88*, 1291–1297. [CrossRef]

19. Hsu, C.-T.; Yao, D.-J.; Ye, K.-J.; Yu, B. Renewable energy of waste heat recovery system for automobiles. *J. Renew. Sustain. Energy* **2010**, *2*, 13105. [CrossRef]

20. Kim, S.-K.; Won, B.-C.; Rhi, S.-H.; Kim, S.-H.; Yoo, J.-H.; Jang, J.-C. Thermoelectric Power Generation System for Future Hybrid Vehicles Using Hot Exhaust Gas. *J. Electron. Mater.* **2011**, *40*, 778–783. [CrossRef]

21. Antonova, E.; Looman, D. Finite elements for thermoelectric device analysis in ANSYS. In Proceedings of the ICT 2005 24th International Conference on Thermoelectrics, Institute of Electrical and Electronics Engineers (IEEE), Clemson, SC, USA, 19–23 June 2005; pp. 215–218.

22. Rezaniakolaie, A.; Rosendahl, L. A comparison of micro-structured flat-plate and cross-cut heat sinks for thermoelectric generation application. *Energy Convers. Manag.* **2015**, *101*, 730–737. [CrossRef]

23. Weng, C.-C.; Huang, M.-J. A simulation study of automotive waste heat recovery using a thermoelectric power generator. *Int. J. Therm. Sci.* **2013**, *71*, 302–309. [CrossRef]

24. Martinez, A.; Vián, J.G.; Astrain, D.; Rodriguez, A.; Berrio, I. Optimization of the Heat Exchangers of a Thermoelectric Generation System. *J. Electron. Mater.* **2010**, *39*, 1463–1468. [CrossRef]

25. Tang, Z.B.; Deng, Y.D.; Su, C.Q.; Yuan, X.H. Fluid Analysis and Improved Structure of an ATEG Heat Exchanger Based on Computational Fluid Dynamics. *J. Electron. Mater.* **2014**, *44*, 1554–1561. [CrossRef]

26. Su, C.Q.; Xu, M.; Wang, W.S.; Deng, Y.D.; Liu, X.; Tang, Z.B. Optimization of Cooling Unit Design for Automotive Exhaust-Based Thermoelectric Generators. *J. Electron. Mater.* **2015**, *44*, 1876–1883. [CrossRef]

27. Wang, Y.; Wu, C.; Tang, Z.-B.; Yang, X.; Deng, Y.; Su, C. Optimization of Fin Distribution to Improve the Temperature Uniformity of a Heat Exchanger in a Thermoelectric Generator. *J. Electron. Mater.* **2014**, *44*, 1724–1732. [CrossRef]

28. Deng, Y.D.; Zhang, Y.; Su, C.Q. Modular Analysis of Automobile Exhaust Thermoelectric Power Generation System. *J. Electron. Mater.* **2014**, *44*, 1491–1497. [CrossRef]

29. Yu, S.; Du, Q.; Diao, H.; Shu, G.; Jiao, K. Start-up modes of thermoelectric generator based on vehicle exhaust waste heat recovery. *Appl. Energy* **2015**, *138*, 276–290. [CrossRef]

30. Högblom, O.; Andersson, R. A simulation framework for prediction of thermoelectric generator system performance. *Appl. Energy* **2016**, *180*, 472–482. [CrossRef]

31. Högblom, O.; Andersson, R. Analysis of Thermoelectric Generator Performance by Use of Simulations and Experiments. *J. Electron. Mater.* **2014**, *43*, 2247–2254. [CrossRef]

32. Aranguren, P.; Astrain, D.; Rodriguez, A.; Martinez, A. Experimental investigation of the applicability of a thermoelectric generator to recover waste heat from a combustion chamber. *Appl. Energy* **2015**, *152*, 121–130. [CrossRef]

33. Jang, J.-C.; Chi, R.-G.; Rhi, S.-H.; Lee, K.-B.; Hwang, H.-C.; Lee, J.-S.; Lee, W.-H. Heat Pipe-Assisted Thermoelectric Power Generation Technology for Waste Heat Recovery. *J. Electron. Mater.* **2015**, *44*, 2039–2047. [CrossRef]

34. He, W.; Su, Y.; Wang, Y.; Riffat, S.; Ji, J. A study on incorporation of thermoelectric modules with evacuated-tube heat-pipe solar collectors. *Renew. Energy* **2012**, *37*, 142–149. [CrossRef]

35. Andersson, B.; Andersson, R.; Håkansson, L.; Mortensen, M.; Sudiyo, R.; van Wachem, B.G.M. *Computational Fluid Dynamics for Engineers*; Cambridge University Press: Cambridge, UK, 2011.

36. Menter, F.R. Two-equation eddy-viscosity turbulence models for engineering applications. *AIAA J.* **1994**, *32*, 1598–1605. [CrossRef]

37. Modest, M.F. *Radiative Heat Transfer*; Elsevier BV: Amsterdam, The Netherlands, 2003.

38. Montecucco, A.; Siviter, J.; Knox, A.R. The effect of temperature mismatch on thermoelectric generators electrically connected in series and parallel. *Appl. Energy* **2014**, *123*, 47–54. [CrossRef]

39. Launder, B.E.; Sharma, B.I. Application of the energy-dissipation model of turbulence to the calculation of flow near a spinning disc. *Lett. Heat Mass Transf.* **1974**, *1*, 131–137. [CrossRef]

40. Thayer, J. Analysis of a heat pipe assisted heat sink. In Proceedings of the 9th International FLOTHERM Users Conference, Orlando, FL, USA, 18–19 October 2000.

41. Chung, D. Materials for thermal conduction. *Appl. Therm. Eng.* **2001**, *21*, 1593–1605. [CrossRef]

42. Brohez, S.; Delvosalle, C.; Marlair, G. A two-thermocouples probe for radiation corrections of measured temperatures in compartment fires. *Fire Saf. J.* **2004**, *39*, 399–411. [CrossRef]

Article

An Experimental Study on the Fire Spread Rate and Separation Distance between Facing Stores in Passage-Type Traditional Markets

Hong-Seok Yun [1], Dong-Gun Nam [2] and Cheol-Hong Hwang [1,*]

[1] Department of Fire and Disaster Prevention, Daejeon University, 62 Daehak-ro, Dong-Gu, Daejeon 34520, Korea; fayayun@gmail.com
[2] Department of Research and Development Laboratory, Korea Fire Institute, 331 Jisam-ro, Giheung-Gu, Yongin 17088, Korea; nam@kfi.or.kr
* Correspondence: chehwang@dju.ac.kr; Tel.: +82-42-280-2592

Received: 28 July 2020; Accepted: 27 August 2020; Published: 28 August 2020

Abstract: Real-scale fire experiments were conducted to understand the fire spread characteristics of the major combustibles handled in traditional markets, a space with high fire risk. The major combustibles were selected through field surveys administered at a number of traditional markets. Through real-scale fire experiments, the horizontal fire spread rate according to the maximum heat release rate of major combustibles was examined. In addition, the separation distance to prevent fire spread to the facing store by radiant heat transfer was examined. As a result of the experiments, it was confirmed that the arrangement method of the combustibles causes a large change in the maximum heat release rate, fire growth rate, and fire spread rate. The horizontal fire spread rate showed a linear proportional relationship with respect to the maximum heat release rate regardless of the type of combustibles, and a correlation to define the relationship was proposed. A correlation equation for predicting the separation distance that can prevent fire spread by radiant heat transfer was proposed, and the curve by the correlation equation was in good agreement with the experimental results. Through this study, it is expected that the correlation proposed to examine the horizontal fire spread rate and the separation distance of major combustibles in a traditional market can be usefully used in the design of fire protection systems to reduce fire damage in the traditional market.

Keywords: traditional market; maximum heat release rate; fire spread rate; radiant heat flux; separation distance

---

1. Introduction

In the field of fire engineering, the use of simulation programs for fire risk assessment of buildings is increasing. The most commonly used Fire Dynamics Simulator (FDS) [1] predicts fire phenomena according to the input parameters given by the user, so the accuracy of the input parameters is very important. This has increased the importance of real-scale fire experiments in the field of fire engineering. Recently, studies for securing input parameters and understanding fire phenomena through real-scale fire tests according to the use of buildings have been actively conducted. For instance, Madrzykowski and Johnsson [2] and Bwalya et al. [3] conducted real-scale fire experiments in residential spaces and presented a measurement of the heat release rate according to the sprinklers' operation. Bennetts et al. [4] conducted office fire experiments to measure temperature, heat flux, and carbon monoxide (CO) concentration related to human safety standards. In addition, Zalok et al. [5] and Hadjisophocleous et al. [6] proposed a fuel package representing fires in commercial premises based on fire loads and present design fire through fire experiments for the fuel package. In another study, Gross et al. [7] presented fire severity according to fire load in hospitals, schools, and warehouses, including residential spaces, to quantify the fire scale. These studies have significantly contributed to ensuring buildings' fire safety in modern society. However, most studies only consider modern buildings and spaces with high fire risk due to outdated facilities not being considered.

Most traditional markets were formed naturally before modern sales and distribution facilities were established, and can be found all over the world. Traditional markets are not systematically designed; therefore, it is composed of temporary buildings and tends to have low fire resistance. Moreover, the indiscriminate use of gas and electric appliances is a factor that increases traditional markets' fire risk. In fact, many fire accidents still occur in Southeast Asian countries where traditional markets have developed. For example, a fire in a traditional market in South Korea burned down about 700 stores and caused 40 million dollars' worth of property damage [8]. Due to the characteristics of the traditional market used by a large number of unspecified people, it is essential to ensure fire safety because many casualties are caused by fire accident. Researchers in South Korea recently proposed the application of the afforestation system [9] and water-based fire protection system [10,11] to reduce fire damage in traditional markets. However, a clear understanding of fire phenomena in traditional markets is necessary in order to determine the required performance of potential safety measures. However, research focusing specifically on fire characteristics in these markets is difficult to find both in South Korea, which has a high interest in traditional market fires, and in the world. Therefore, in this study, a real-scale fire test was conducted as part of a project for the development of a water curtain system optimized for a traditional market and fire protection design throughout. For this, major combustibles with high fire loads were selected in the traditional market. In the experiment, the heat release rate, the incident heat flux according to the distance were measured, and the fire spreading phenomena was examined.

Most traditional markets are designed as "passage-type", with stores facing each other across a passageway. Stores in the passage-type traditional market are separated by simple structures, such as thin walls or sandwich panels. As a result, fire can spread very easily between adjacent stores via direct flame contact. Since the fire spread is closely related to an increase in the fire growth rate—or the maximum heat release rate according to combustion area—understanding the fire spread characteristics is very important in establishing a fire safety plan. Most of these studies analyze the fire spread phenomena in a well-controlled thermal environment targeting widely used materials [12–14]. However, in actual fire environments, the fire spread can vary due to numerous factors, such as the amount, arrangement, and location of combustibles. As a result, defining the fire spread through bench-scale experiments is sometimes a limited approach. Recently, research has been conducted to predict the fire spread of combustibles through simulation [15,16]. However, prediction of thermal decomposition and fire spread phenomena of combustibles through simulation requires the input of numerous factors, such as activation energy, pre-exponential factor, and heat of combustion (including thermal properties). Since these values must be measured very precisely and are required individually for each material, there are limitations in predicting the actual fire spread at the current stage of research. Therefore, real-scale fire testing on target combustibles is the most effective way to understand fire spread characteristics. Additionally, examining the relationship between the physical quantities obtained from the experiment and the fire spread rate can be used as a simple means to replace various and complicated input parameters required in simulation.

In addition to the fire spread to adjacent stores by direct flame contact, the fire can spread to facing stores with passages in between. The fire on the facing store, i.e., on the new line, spreads to adjacent stores located on the same line by flame contact, which can greatly increase the size of the fire. Therefore, a number of studies have been conducted on the necessary separation distance to reduce damage caused by fire spreading across separate spaces [17–22]. These studies mainly considered the vertical and horizontal fire spread between buildings due to contact of the facade flame and flying brand. However, in the traditional market, facing stores are not separated by walls, and combustibles placed in the passages are very close, so they have a relatively high fire risk. Therefore, examining the separation distance suitable for the fire environment of the traditional market is requried. Radiant heat flux is usually what causes fire to spread between stores facing each other in passage-type traditional markets. Radiant heat flux can cause thermal decomposition and ignition of combustibles located far away without direct flame contact. Thus, identifying the precise separation distance where the radiant heat flux emitted from the fire source decreases below the critical value can provide useful information for reducing fire damage in traditional markets.

Based on this background, in this study, real-scale fire experiments were conducted on major combustibles in the traditional market. Major combustibles based on fire load were selected through field surveys of three traditional markets. Clothing, blankets, plastics, and bags were selected as the major combustibles. Experiments and analysis were conducted to quantify the fire spread of combustibles arranged to have a shape and fire load similar to those of the actual traditional market. The relationship between the maximum heat release rate and the fire spread rate was examined, and the distance between the stores to prevent fire spread was suggested through the measurement of the incident radiated heat flux. Finally, the curve based on the theory of incident heat flux for the remote target proposed by Modak [23] and the separation distance derived through experiments were compared. It is expected that the results of this study can be used as basic information useful in the application of water curtain system to prevent fire spread and reduce damage in traditional markets.

2. Fire Characteristics of Traditional Market

2.1. Spreading Paths of the Fire

Figure 1a presents a picture of a passage-type traditional market taken during the field survey process, and the field survey process is described later. Facing stores are arranged in parallel, and these structures are generally similar. The width of the passage varies slightly from market to market, but is generally about 3.0 m. However, as shown in Figure 1a, stalls placed in the passage reduce the width of the passage and the distance between combustibles. Due to the characteristics of these traditional markets, the fire spread, as shown in Figure 1b, may occur. As mentioned earlier, the fire probability of passage-type traditional markets can be classified into two types. Horizontal fire spread refers to the fire spread along the combustible materials of a store wall or stalls via direct contact with the flames. Since this type of fire spread is limited to stores along the same line, damage can be reduced through water curtain systems or fire shutters considering the fire spread rate. Implementing these systems may also be effective in preventing the fire spread in facing stores via radiant heat flux. However, from a conservative point of view, it may be effective to ensure a minimum distance (r) in which the radiant heat flux—which is emitted from a fire source and reaching the opposite store is reduced below a critical value. Therefore, in order to ensure safety by preventing the fire spread in traditional markets, it is necessary to assess the fire spread rate and separation distance of major combustibles through experiments.

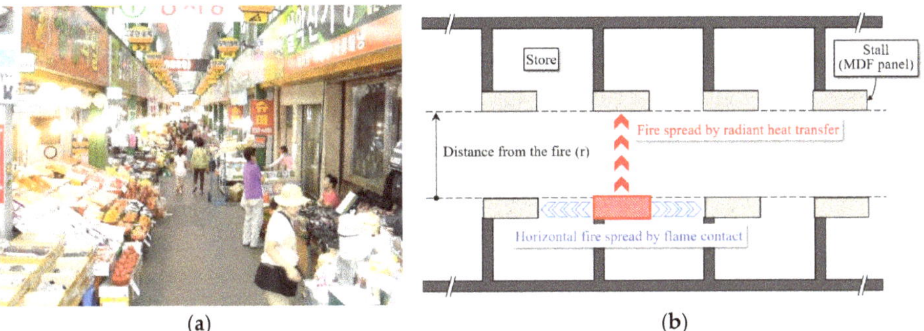

(a)   (b)

Figure 1. General structure of the passage-type traditional markets and fire spread path: (**a**) Photo of passage-type traditional market; (**b**) fire spread path in the traditional markets.

2.2. Field Survey to Select Major Combustibles

Figure 2 presents a comparison of fire loads based on store type for three traditional markets in South Korea. Different store types were classified based on their items and goods. Specifically, 11 types of items are most commonly handled in traditional markets, including clothing, plastics, vinyl flooring, blankets, bags, cosmetics, shoes, hats, and agricultural and fisheries products. In order to derive the fire load (MJ/m^2) of each store, the mass of all combustibles (kg) in the store was measured, and the total amount of heat was calculated from the heat of combustion (kJ/kg) obtained through Thermal gravimetric analysis (TGA) [24] and literature review [25]. After that, the fire load was derived by considering the relationship between the store floor area (m^2) and the total amount of heat. Finally, in the figure, the average value of fire load for each store type obtained in three traditional markets and the standard deviation for this are presented together. Based on the study's comparison of fire loads by store type, clothing stores had the highest fire load at approximately 650 MJ/m^2. The next highest fire load of approximately 530 MJ/m^2 belonged to stores handling plastics and vinyl flooring. This was then followed by blankets, bags, cosmetics, shoes, hats, and agricultural and fisheries products. Taking fire load into account, four major combustibles were selected. Clothing and plastics with high fire loads were selected first, while vinyl flooring with fire loads similar to those of plastics were excluded. Blankets and bags with high fire loads were also selected as major combustibles.

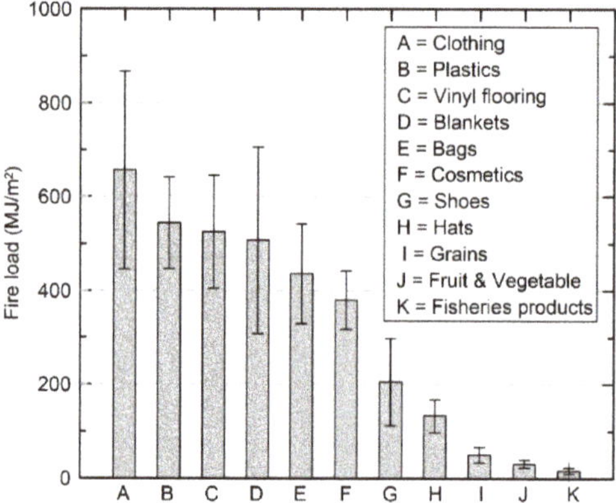

Figure 2. Comparison of fire loads by store type through field survey to select major combustibles in traditional markets.

3. Experimental Details

Figure 3a to e indicate the arrangement of the major combustibles on hangers and stalls for real-scale fire testing. Figure 3a,b indicate the arrangement of clothing, with clothing types including bubble jackets, pants, and t-shirts. In actual stores, clothing is typically displayed on hangers or stacked on sales stalls; therefore, two layout methods were employed to reflect this. Figure 3a shows the clothing hanging on eight hangers with a total mass of 79.2 kg. Figure 3b shows clothing uniformly distributed on two stalls with a total mass of 63.6 kg. The hangers and stalls used in this study had the same shapes as those most commonly used in traditional markets. Each stall features a 1.65 m × 0.82 m area and 25 mm thick MDF panel on top of a steel support fixture. Figure 3c displays the layout chosen for blankets, with nine thick blankets and 21 thin blankets laid over two stalls. The total combustible mass for blankets was 68.0 kg. Figure 3d displays the layout of plastics, for which only one stall was used. In detail the types of combustibles include trash cans, covered side dishes cases and laundry baskets. The total mass of combustibles was 22.3 kg. Figure 3e shows a photo of bags, and products of various materials such as synthetic leather, acrylonitrile butadiene styrene (ABS), polyethylene, and polycarbonate were placed on two stalls. The total combustible mass of bags was 30.0 kg. In all experimental conditions, the fire load according to the floor area occupied by the hangers and stalls, the total mass of the combustibles was set to have a value within the deviation of the average value identified through field survey. In experiments with major combustibles, the heat release rate and incident radiant heat flux were measured. To measure the heat release rate, a large-scale calorimeter (LSC) was used, and all experiments were conducted under an exhaust hood with a diameter of 10.0 m. For the measurement of the incident radiant heat flux, a plate-thermometer (PT) was used. The PT was developed by Wickström [26] and registered as an international standard to review the fire resistance performance of building materials inside furnaces in the 1980s [27]. However, recently, the heat transfer equation is inverted as shown in Equation (1) for the temperature measured from the rear center, and PT is used to measure the incident heat flux [28–30]. PT can be used as a suitable alternative to heat flux meters that require various additional devices such as purging and water cooling systems to measure heat flux.

$$\left[\dot{q}''_{inc}\right]^{i+1} = \sigma\left[T^4_{PT}\right]^i + \frac{K_{cond}\left([T_{PT}]^i - T_\infty\right) + h_{PT}\left([T_{PT}]^i - \left[T_g\right]^i\right) + \rho_{st}C_{st}\delta\frac{[T_{PT}]^{i+1} - [T_{PT}]^i}{[t]^{i+1} - [t]^i}}{\varepsilon_{PT}} \tag{1}$$

where $\dot{q}''_{inc}$ represents the incident heat flux measured by PT. σ, K_{cond}, h_{PT} stand for Stefan-Boltzmann constant, conduction correction factor (8.6 W/m^2·K) [31], convective heat transfer coefficient (10 W/m^2·K). T_{PT}, T_g, T_∞ refer to the PT's surface temperature, surrounding temperature, and ambient temperature. In addition, ρ_{st}, C_{st}, δ are the properties of the metal plate related to thermal inertia, and are density (8000 kg/m^3), specific heat (477 J/kg·K), and thickness (0.6 mm).

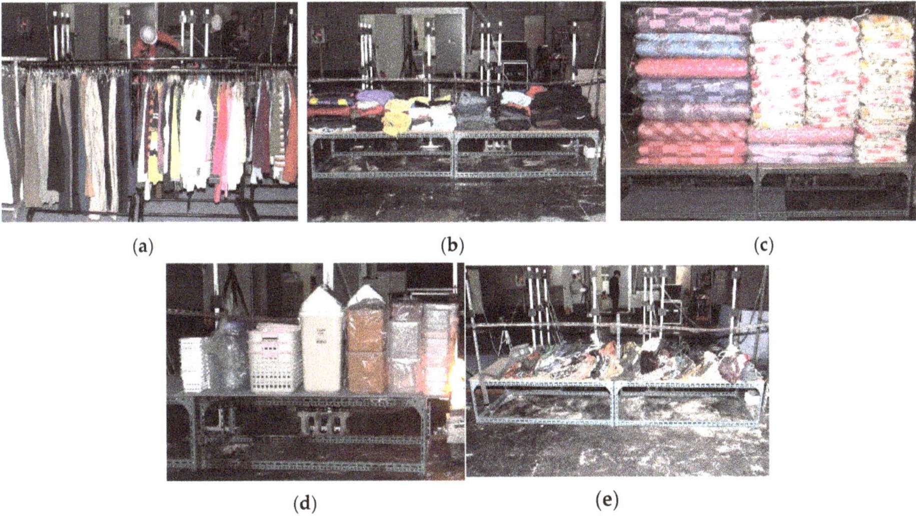

(a)   (b)   (c)

(d)   (e)

Figure 3. Photographs of the major combustibles arranged to conduct real-scale fire experiments: (**a**) Clothing hanging on the hangers; (**b**) clothing placed on the stalls; (**c**) blankets placed on the stalls; (**d**) plastics placed on the stall; (**e**) bags placed on the stalls.

Table 1 presents a summary of the experimental conditions such as total mass and heat of combustion of the combustibles and fire loads. The heat of combustion for each combustibles was extracted from the results of a prior study [24] and from the Fire Protection Handbook [25] published by the National Fire Protection Association (NFPA). Furthermore, in the case of bags, an average value considering the bags' various materials used as combustibles was applied.

Table 1. Summary of experimental conditions of major combustibles considered in the present study.

Combustibles	Total Mass (kg)	Average Mass Per Unit Area (kg/m^2)	Heat of Combustion (MJ/kg)	Fire Load (MJ/m^2)	
				Experiment	Field Survey
Clothing (hanger)	79.19	18.45	23.30	430.09	656.39 ± 211.0
Clothing (stall)	63.56	23.49		547.28	
Blankets	68.00	25.13	17.41	437.50	507.48 ± 198.9
Plastics	22.29	16.47	34.33	565.57	543.90 ± 97.5
Bags	30.01	11.09	30.73	340.80	436.16 ± 106.0

Figure 4a,b show the layout of hangers, stalls and measuring devices for fire experiment of major combustibles. In Figure 4a, eight hangers arranged in four columns occupy a floor area of 3.9 m (x) × 1.1 m (y), and column and row information to describe the location of the hanger and measuring device are also indicated. A burner for ignition was placed between the two hangers on the first column. In order to provide the only minimum ignition energy, the size of the burner and amount of heptane were determined to be 0.1 m (x) × 0.1 m (y) × 0.04 m (z) and 45 mL through several tests. In this case, the heat release rate of the burner is about 8.2 kW and the fire duration time is 180 s. Nine PT stands were arranged in three rows to measure the maximum incident heat flux according to distance from combustibles. When, considering the concept of the view factor, the tilt angle between the fire source and the receive surface will affect the incident heat flux. In the process of fire spread along the combustibles, the tilt angle between the burning location and the PT can cause errors in the measurement of the incident heat flux. In order to minimize the possibility of such errors, five PT stands were placed at 0.9 m intervals in the first row closest to the combustibles. The second and third rows were placed further apart at intervals of 0.7 m each. In general, it can be predicted that the influence of the tilt angle (or view angle) will decrease as the distance from the heat source increases. Therefore, only two stands were placed in second and third rows. The PT stands in the second and third row were moved horizontally by 0.1 m to prevent shielding by the stand located in the front row. Each stand has a PT installed at a height of 1.0 m and 1.5 m from the floor. Figure 4b shows a layout for experiments on combustibles (clothing, blankets, plastics, and bags) placed on a stalls. The floor area occupied by the two stalls is 3.3 m (x) × 0.82 m (y). In the case of plastics, only one stall was used (The fire loads presented in Table 1 were calculated according to the number of stalls used). The shape of the burner and amount of heptane were set the same as described in Figure 4a. Relatively short flame length compared to the height of the stall (0.5 m) was insufficient to ignite combustibles. So the burner was installed at a height of 0.24 m from the floor. Combustibles placed on the stalls were expected to have a smaller fire scale than clothing hanging on hangers having a vertical arrangement. Therefore, PT stands were moved by 0.2 m towards the combustibles. Under these conditions, the heat release rate, fire growth rate, and incident heat flux of each combustibles were measured and compared. In all experiments, the amount of combustibles placed on each hanger or stall was set evenly in terms of mass. In addition, video recording was performed throughout the experiments, and the fire spread rate was analyzed through the video analysis.

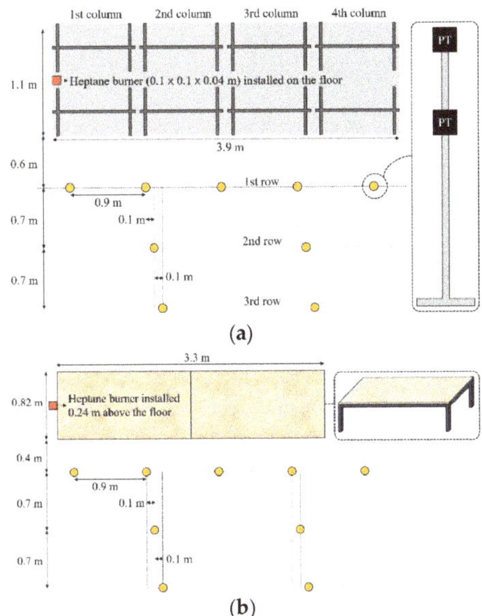

Figure 4. Layout of plate-thermometer stands for measuring the incident radiant heat flux of major combustibles: (**a**) Experiment with clothing hanging on hangers; (**b**) experiment on combustibles placed on the stalls.

4. Results and Discussion

4.1. Relationship between the Maximum Heat Release Rate and Fire Spread Rate

The heat release rate was measured through LSC using the real-scale fire experiments with major combustibles in traditional markets. In all experiments, measurements were taken until all of the combustibles had burned down. However, measurements are only presented for sections where significant variations in heat release rates were identified in order to examine the phenomena of early fire growth and fire peak periods in detail. Figure 5a,b shows photos of the combustion process over time and results of the heat release rate measurements during the experiment with clothing hanging on hangers, respectively. Figure 5a illustrates the fire spread pattern over time from the ignition of burners. At around 3 min, the fire spread was limited to two hangers located in the first column. The fire spread followed a mostly vertical pattern due to the arrangement of the clothing. In the photo taken at 5 min after the hangers of the first column collapsed, it can be seen that the fire spread from the clothing burning on the floor to the combustibles placed on the second column. Similarly, the fire spread to combustibles located in third column was initiated after the hangers in second column collapsed, as shown in the photo about 8 min. The same trend occurred when the fire spread to the fourth column. In the case of clothing hanging on the hangers, the fire spread was mainly affected by the collapse of the hanger, and it was confirmed that the spread of vertical fires occurring in each column influenced the fire growth. Figure 5b shows that throughout the combustion process of clothing, the heat release rate exhibited four peaks with very similar patterns and a maximum heat release rate value of approximately 2300 kW. This can be interpreted as the combustion of each column's combustibles. The heat release rate indicates the peak value and then a reduction due to the exhaustion of combustibles until the fire spread to the next column's hanger. However, after the fire spread to the next column, vertical fire spread repeatedly contributed to the fire growth. The fire growth rate of all peak calculated by time-square law [32] is 'Ultra-fast' ($\alpha = 0.1874$ kW/s^2). An analysis of the video confirmed that the horizontal fire spread rate was approximately 0.325 m/min.

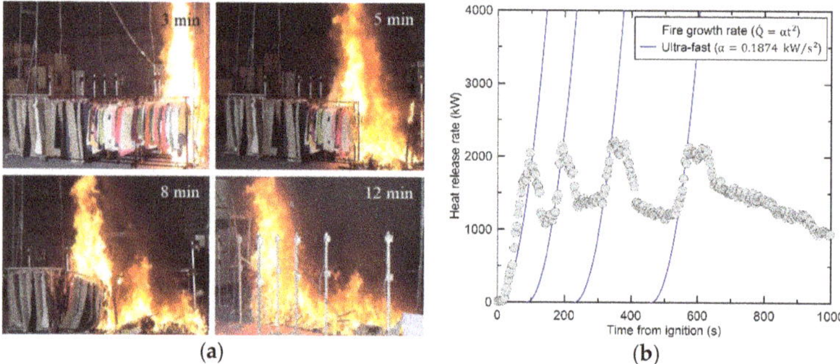

Figure 5. Fire spread pattern and heat release rate of clothing hanging on hangers: (**a**) Sequential photographs of fire spread over time; (**b**) measured heat release rate.

Figure 6 presents the results of another experiment where the same combustibles as shown in Figure 5 was placed on stalls. Figure 6a illustrates the fire spread over time, which started at three minutes on clothing closest to the burner. The flames then spread horizontally along the combustibles located at the top. At the 10-min mark, the flame spread throughout the first stall. At the 20-min mark, fire spread to all combustibles used in the experiment. At around 32 min, the combustibles fell to the floor due to the combustion of the MDF panel located on the first stall (confirmed by the photos at 35 min). Figure 6b presents the results of the heat release rate measurement over time. The initial fire growth rate was 'Slow' ($\alpha = 0.0029$ kW/s^2), which was significantly lower than the rate of clothing hanging on hangers despite the combustibles being the same. The maximum heat release rate occurred at 1100 s ($\approx$20 min) and was approximately 1300 kW. After this peak period, the size of the fire was gradually reduced by the exhaustion of combustibles. A 'medium' fire growth rate ($\alpha = 0.0117$ kW/s^2) was measured at 1900 s ($\approx$32 min), which is due to collapse of the MDF panel. The horizontal fire spread rate for clothing placed on a stalls was 0.178 m/min, slower than that of clothing hanging on hangers. The significant difference in fire spread rate for same combustibles is due to difference in fire spread process. In the experiment of clothing placed on stalls, fire spread occurring horizontally along the combustibles is observed. However, for clothing hanging on hangers, fire spread by collapse of hangers is observed. In other words, it is difficult to directly compare the fire spread rate according to the arrangement method of combustibles. However, differences in the maximum heat release rate and fire growth rate can be confirmed. The maximum heat release rate of clothing increases by about 1.8 times depending on the arrangement method, and the fire growth rate also shows a large difference from 'slow' to 'ultra-fast'.

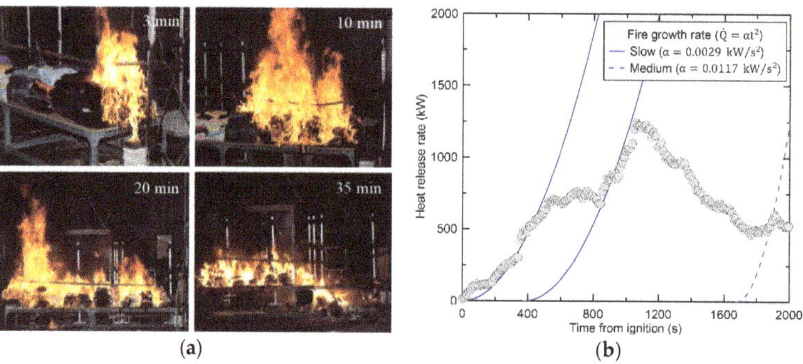

Figure 6. Fire spread pattern and heat release rate of clothing placed on stalls: (**a**) Sequential photographs of fire spread over time; (**b**) measured heat release rate.

Figure 7 represents the results of an experiment where blankets were placed on stalls. Figure 7a is a sequential photographs of the fire spread pattern over time. As can be seen in the Figure 7a, combustion did not occur until 2 min after the burner was ignited. However, as the blankets' plastic bags burned, the fire spread rapidly to the top of the combustibles in about 150 s (2 min 30 s). The fire initially exhibited a vertical spread, but it then spread horizontally along the combustible material (between 4 and 7 min). The fire spread to almost all combustibles at around 9 min, and it then spread out completely at around 11 min 30 s. Figure 7b presents the results of the heat release rate measurement. The first fire growth occurred before reaching 200 s from ignition of the burner, which is the result of vertical fire spread from the plastic bags. The fire growth rate via horizontal fire spread is also shown, indicating a 'Fast' fire growth rate (α = 0.0468 kW/s^2). At the peak period, the maximum heat release rate exhibited a value of approximately 3500 kW. The maximum heat release rate appeared before the fire spread was completed. The fire was extinguished through decay following exhaustion of the combustibles. A temporary 'medium' fire growth rate (α = 0.0117 kW/s^2) occurred at approximately 900 s, the cause of which was confirmed to be the collapse of the MDF panel. Horizontal fire spread rate of the blankets as identified by the video was 0.287 m/min.

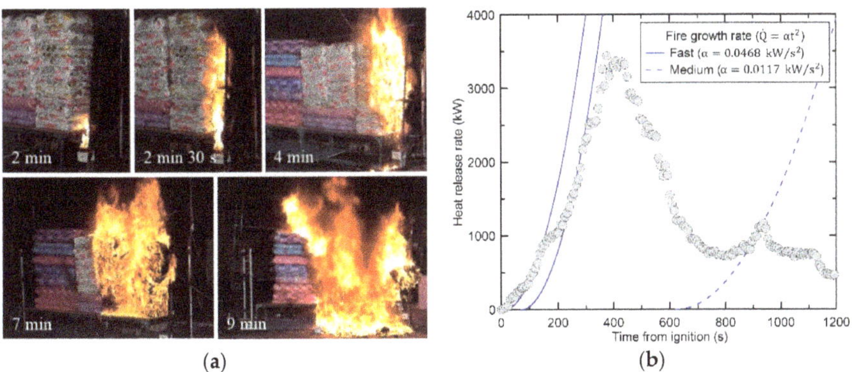

Figure 7. Fire spread pattern and heat release rate of blankets placed on stalls: (**a**) Sequential photographs of fire spread over time; (**b**) measured heat release rate.

Figure 8 presents the results of the heat release rate measurement obtained from the experiments with plastics and bags. With these combustibles, no particular phenomena were observed during the fire spread; thus, no photographs are presented. Based on the heat release rate of plastic products illustrated by Figure 8a, no fire growth was achieved until 600 s. This is due to the nature of thermoplastic resin, which burns after being melted by heat. The fire growth rate during combustion of molten combustibles showed a value of 'medium' (α = 0.0117 kW/s^2). The fire, which grew continuously, was extinguished after exhibiting a maximum heat release rate of 2200 kW at its peak period. It took 570 s for the fire to spread to all combustibles from the initial moment of combustion, and the horizontal fire spread rate was 0.174 m/min. The fire growth rate of the bags confirmed through the experiment showed a very low value of 'Slow' (α = 0.0029 kW/s^2) as shown in Figure 8b. In addition, the maximum heat release rate was about 600 kW, and the bags showed the lowest maximum heat release rate among the combustible materials considered. A period of temporary fire growth occurred at around 2500 s from ignition, caused by the collapse of the MDF panel placed on the stall. The horizontal fire spread rate of the bags placed on the stalls was identified by the video to be 0.120 m/min.

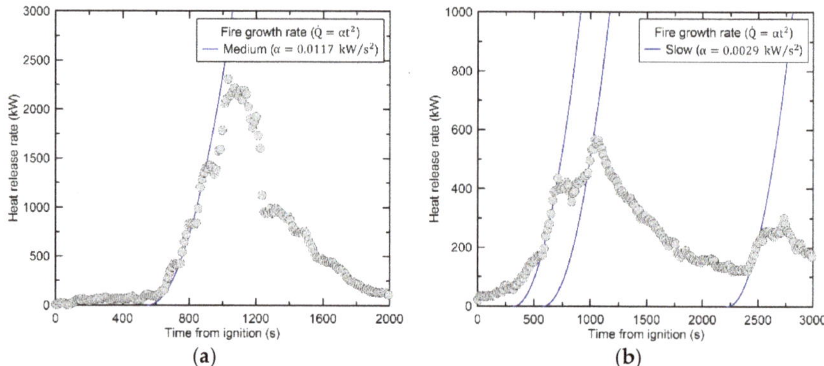

Figure 8. Measured heat release rate and fire growth curve: (**a**) Plastics; (**b**) bags.

Table 2 presents a summary of the experimental results. The fire growth rate shows a wide distribution from 'slow' to 'ultra-fast'. The fire growth rate of 'Ultra-fast' was found in the experiment of clothing hanging on hangers, where the tendency of vertical fire spread within each column was dominant. On the other hand, in the experiment of clothing placed on stalls where the influence of horizontal fire spread is dominant is 'slow', showing a significant difference in fire growth rate. In addition, the change in the arrangement method caused a difference of about 1.8 times in the maximum heat release rate. From this, it can be seen that the change in the arrangement method causes a significant change in the fire growth of the same combustibles. Unfortunately, the effect of the arrangement method on the horizontal fire spread rate was not examined in the present study. This is because, as mentioned above, collapse of the hanger affects the horizontal fire spread rate. Therefore, it is difficult to directly compare with experiments in which fire growth is performed by horizontal fire spread along combustibles. However, it is confirmed that the horizontal fire spread rate of other four experiments has a proportional relationship to the maximum heat release rate. Further, the combustibles for the experiments were designed to have similar fire load as in the actual traditional market. When considering this, examining the relationship between the maximum heat release rate and the horizontal fire spread rate can be useful in evaluating the fire spread rate in the traditional market.

Table 2. Summary of experimental results for the major combustibles in traditional markets.

Combustibles	Fire Growth Rate ($\dot{Q} = \alpha t^2$)	Maximum Heat Release Rate (kW)	Horizontal Fire Spread Rate (m/min)
Clothing (hanger)	Ultra-fast ($\alpha = 0.1874$ kW/s^2)	2360	0.325
Clothing (stall)	Slow ($\alpha = 0.0029$ kW/s^2)	1340	0.178
Blankets	Fast ($\alpha = 0.0468$ kW/s^2)	3514	0.287
Plastics	Medium ($\alpha = 0.0117$ kW/s^2)	2405	0.174
Bags	Slow ($\alpha = 0.0029$ kW/s^2)	592	0.120

Figure 9 illustrates the relationship between the maximum heat release rate and the horizontal fire spread rate obtained through experiments. At this time, it was confirmed that the horizontal fire spread rate was affected by the collapse of the hanger, not the burning rate of the combustibles, so the experimental results of the clothing hanging on hangers were excluded. As a result, it can be seen that the horizontal fire spread rate of each combustibles has a proportional relationship to the maximum heat release rate. This relationship between the maximum heat release rate and the fire spread rate can be expressed through Equation (2):

$$\text{Horizontal fire spread rate } [\text{m/min}] = \dot{Q}_{max}\left(5.06 \times 10^{-5}\right) + 0.09 \tag{2}$$

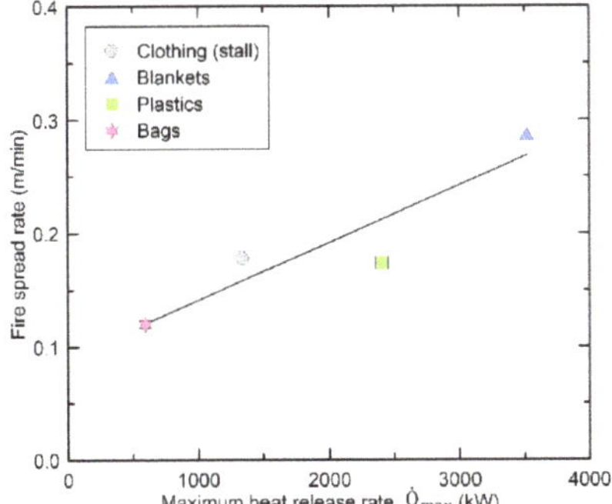

Figure 9. Relationship between the maximum heat release rate and horizontal fire spread rate.

The straight line by Equation (2) shows appropriate agreement with horizontal fire spread rate of the four experiments. Therefore, it can be seen that the maximum heat release rate can be utilized as a simple means to replace various and complex parameters for predicting the fire spread rate. However, the correlation predicts the fire spreading phenomenon only through the relationship with the heat release rate, and it is only examined under the conditions where horizontal fire spread is dominant, so caution is required in use of the correlation.

4.2. Separation Distance to Prevent Fire Spread by Radiant Heat Flux

As mentioned earlier, the fire spread via radiant heat flux in passage-type traditional markets lined with stores facing each other can be fatal. In this study, incident radiant heat flux according to the distance from combustibles was measured to derive the separation distance to prevent fire spread by radiant heat flux. Figure 10 present the maximum value of the incident radiant heat flux measured at various distances through PT in the experiments. Figure 10 shows the maximum value of the incident radiant heat flux measured with distance in each experiment. In addition, a curve for predicting the incident radiant heat flux according to the distance is also presented. The second-order function was used for the curve considering the characteristic of radiant heat flux which is inversely proportional to the square of the distance from the heat source. At this time, the cross point of the curve and the dotted line indicating the critical heat flux can be regarded as the minimum distance that must be guaranteed to prevent the fire spread by the radiant heat flux. The critical heat flux was set to 20 kW/m^2 [33], which is the occurrence condition of flashover caused by radiant heat flux in a general fire environment. In legend of the figure, results of the separation distance according to each experiment are presented. When a distance greater than the required separation distance obtained in each experiment is secured, the incident heat flux will decrease below the critical value. The required separation distance generally tends to increase with the maximum heat release rate. While other experiments show separation distances greater than 1.0 m, clothing and bags placed on stalls show very small separation distances of 0.35 m and 0.01 m. This very small separation distance is due to the PT measurement error. In this study, PT stands were arranged so as not to generate errors according to the continuous flame area and the inclination angle of the measuring surface. That is, a location where the maximum view factor can be secured for the continuous flame area at a given distance was selected. However, in these two experiments, the continuous flame area was formed at a height lower than the installed height (1.0 m and 1.5 m) of the PT. Measurement error due to tilt angle causes inaccuracy of the second-order function, resulting in very small separation distances. That is, it is determined that the incident radiant heat flux actually emitted from the continuous flame area is not accurately measured. However, in other experiments, sufficient flame length (>2.5 m) was ensured, and as a result, PTs were placed in the continuous flame area.

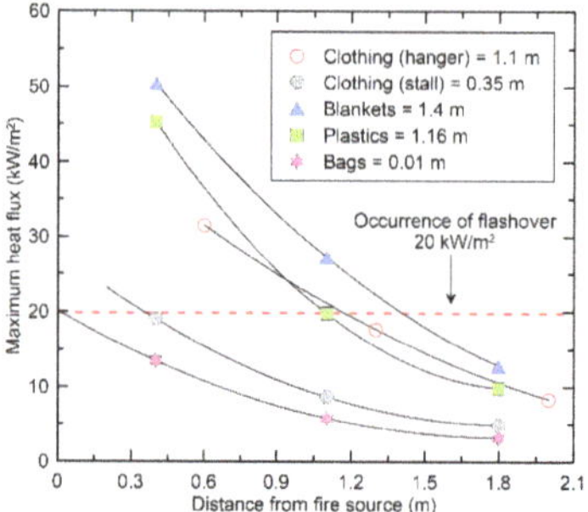

Figure 10. Minimum separation distance to prevent fire spread by radiant heat flux.

The separation distances of major combustibles were examined through experiments, but the information obtained through a few experiments is insufficient to be used for safety design in traditional markets. Therefore, it is required to derive a correlation that can easily evaluate the safety separation distance. For this, Modak's simple method can be used to predict the incident radiant heat flux for the remote target. As is well known, assuming the flame as a cylinder, and considering the view factor, the incident radiant heat flux for the remote target and separation distance can be predicted accurately. However, as confirmed in Figure 1a, there are many difficulties in calculating the separation distance in consideration of the view factor between all combustibles in the traditional market where the store shape is not standardized. Therefore, it may be useful to examine the applicability of the relatively simple Modak's method. The correlation proposed by Modak is expressed as Equation (3) below:

$$\dot{q}'' = \frac{x\dot{Q}_{max}}{4\pi r^2} \tag{3}$$

where $\dot{q}''$, x, $\dot{Q}_{max}$, r refer to the incident heat flux, radiative fraction, maximum heat release rate and distance to the target surface, respectively. If Equation (3) is rearranged for r and 20 kW/m^2 is substituted for $\dot{q}''$, the separation distances according to the maximum heat release rate can be derived. In Figure 11, the mean value of x (0.1516), derived by rearranging Equation (3) for x, has been applied. In Figure 11, the curve agrees very well with the experimental results, except for the clothing and bags placed on the stalls with errors in the measurement of heat flux. In conclusion, it can be seen that Modak's simple method can be used to examine the separation distance for combustibles whose maximum heat release rate and radiative fraction are known. However, when using this method, a suitable measurement location that is not affected by the view factor must be carefully selected to avoid the same error as shown in Figure 10. This is also related to the premise that the target surface should face the heat source in Modak's simple method.

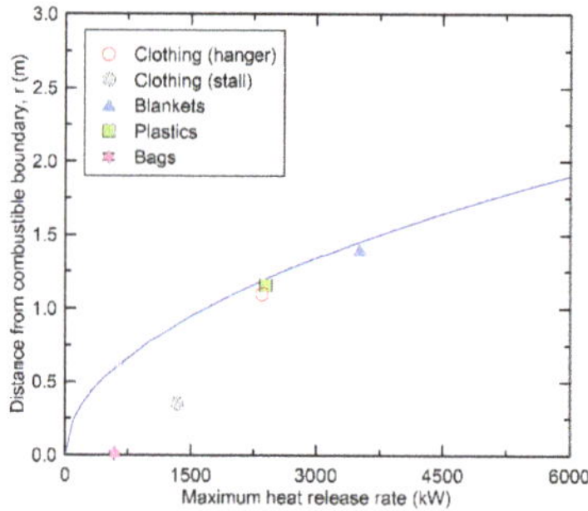

Figure 11. Correlation of separation distance according to maximum heat release rate.

5. Conclusions

In this study, real-scale fire experiments were conducted using major combustibles that are actually present in traditional markets. The major combustibles selected based on the fire load were designed in a shape similar to that existing in the actual traditional market. Through the experiment, the correlation between the horizontal fire spread rate and separation distance according to the maximum heat release rate was derived.

In experiments on the same combustibles (clothing), the maximum heat release rate and fire growth rate varied greatly depending on the arrangement method of the combustibles. Compared with the horizontal arrangement condition, the maximum heat release rate in the vertical arrangement condition increased about 1.8 times, and the fire growth rate increased significantly from 'slow' to 'ultra-fast'. This change is caused by the difference in the fire spread direction according to the arrangement method of combustibles.

The horizontal fire spread rate via flame contact was confirmed to have a linear proportional relationship to the maximum heat release rate regardless of combustible type, and a correlation was derived to define it. The correlation derived in this study can be used as a simple means to replace various and complex parameters for predicting the spread of fire.

The applicability of the Modak's simple method to evaluate of the safety separation distance was evaluated through comparison with the experimental results. The curve obtained from the correlation showed a tendency to agree considerably with the experimental results in which the incident heat flux was properly measured. The use of the correlation is expected to be useful in examining the separation distance to prevent the spread of fire to opposite stores by radiant heat flux in the traditional market.

Author Contributions: Conceptualization: H.-S.Y. and C.-H.H.; methodology: H.-S.Y. and C.-H.H.; software: H.-S.Y. and D.-G.N.; validation: H.-S.Y. and C.-H.H.; formal analysis: H.-S.Y. and C.-H.H.; resources: C.-H.H.; data curation: H.-S.Y. and D.-G.N.; writing—original draft preparation: H.-S.Y.; writing—review and editing: C.-H.H.; visualization: H.-S.Y.; supervision: C.-H.H.; project administration: C.-H.H.; funding acquisition: C.-H.H. All authors have read and agreed to the published version of the manuscript.

Funding: This research was funded by [National Fire Agency of Korea] grant number [2015-NFA001-004] And the [Ministry of Public Safety and Security] grant number [NEMA-Next Generation-2014-49].

Conflicts of Interest: The authors declare no conflict of interest.

References

1. McGrattan, K.; Hostikka, S.; McDermott, R.; Floyd, J.; Vanella, M.; Weinschenk, C.; Overholt, K. *FDS User's Guide*, 6th ed.; NIST Special Publication 1019; National Institute of Standard and Technology: Gaithersburg, MD, USA, 2017.
2. Madrzykowski, D. Impact of a Residential Sprinkler on the Heat Release Rate of a Christmas Tree Fire. USA, 2008. NIST Interagency/Internal Report 7506. Available online: https://tsapps.nist.gov/publication/get_pdf.cfm?pub_id=861535 (accessed on 20 January 2017).
3. Bwalya, A.; Gibbs, E.; Lougheed, G.; Kashef, A. Heat release rate of modern residential furnishings during combustion in a room calorimeter. *Fire Mater.* **2015**, *39*, 685–716. [CrossRef]
4. Bennetts, I.D.; Moinuddin, K.A.; Thomas, I.R.; Proe, D.J. Sprinklered office fire tests. *Fire Mater.* **2008**, *32*, 159–198. [CrossRef]
5. Zalok, E.; Hadjisophocleous, G.V.; Mehaffey, J.R. Fire loads in commercial premises. *Fire Mater.* **2009**, *33*, 63–78. [CrossRef]
6. Hadjisophocleous, G.V.; Zalok, E. Development of Design Fires for Performance-Based Fire Safety Designs. *Fire Saf. Sci.* **2008**, *9*, 63–78. [CrossRef]
7. Gross, D. Measurement of Fire Loads and Calculations of Fire Severity. *Wood Fiber Sci.* **1977**, *9*, 72–85.
8. National Fire Data System. Available online: http://nfds.go.kr/stat/general.do (accessed on 7 May 2020).
9. Oh, J.H.; Kim, D.Y. Fire-Resistant Green-Wall System for Traditional Markets. *J. Archit. Inst. Korea Struct. Constr.* **2017**, *33*, 27–33. [CrossRef]

10. Park, J.W.; Jeon, G.Y.; Na, W.J.; Hong, W.H. A Study on Features of Fire and Change of Extinguishing Capacity of Sprinkler Equipment against Fire at a Traditional Market. *Fire Sci. Eng.* **2008**, *22*, 128–138. Available online: http://kifsejournal.or.kr/upload/pdf/HJSBCY_2008_v22n1_128.pdf (accessed on 11 April 2018).

11. Lee, S.C.; Kim, B.J.; Lee, J.O.; Park, C.H.; Hwang, C.H. An Experimental Study on the Effects of the Shape of a Drencher Head on the Characteristics of a Water Curtain. *Fire Sci. Eng.* **2016**, *30*, 86–93. [CrossRef]

12. Rothermel, R.C.; Anderson, H.E. *Fire Spread Characteristics Determined in the Laboratory*; United States Forest Serviece: Ogden, UT, USA, 1966.

13. Zhang, Y.; Huang, X.; Wang, Q.; Ji, J.; Sun, J.; Yin, Y. Experimental study on the characteristics of horizontal flame spread over XPS surface on plateau. *J. Hazard. Mater.* **2011**, *189*, 34–39. [CrossRef]

14. Zhao, K.; Zhou, X.; Peng, F.; Ju, X.; Yang, L. Experimental Study on Flame Spread over Inclined PMMA and PE Slabs. *Key Eng. Mater.* **2018**, *775*, 390–394. [CrossRef]

15. Chaos, M.; Khan, M.M.; Krishnamoorthy, N.; De Ris, J.L.; DoroFeev, S.B. Evaluation of optimization schemes and determination of solid fuel properties for CFD fire models using bench-scale pyrolysis tests. *Proc. Combust. Inst* **2011**, *33*, 2599–2606. [CrossRef]

16. Park, W.H. Optimization of fire-related properties in layered structures. *J. Mech. Sci. Technol.* **2019**, *33*, 3031–3840. [CrossRef]

17. Tang, F.; Hu, L.H.; Qui, Z.W.; Zhang, X.C.; Lu, K.H. Window ejected flame height and heat flux along façade with air entrainment constraint by sloping facing wall. *Fire Saf. J.* **2015**, *71*, 248–256. [CrossRef]

18. Hu, L.; Hu, K.; Ren, F.; Sun, X. Façade flame height ejected from an opening of fire compartment under external wind. *Fire Saf. J.* **2017**, *92*, 151–158. [CrossRef]

19. Maranghides, A.; Johnsson, E.L. Residential Structure Separation Fire Experiment, USA, 2008. NIST Technical Note 1600. Available online: https://www.govinfo.gov/content/pkg/GOVPUB-C13-61ebeb4a9065a1584c3a9b0561bbb525/pdf/GOVPUB-C13-61ebeb4a9065a1584c3a9b0561bbb525.pdf (accessed on 10 May 2020).

20. Delichatsios, M.A.; Ryan, J.; Tian, N.; Zhang, J. Vertical safe separation distance between openings in multi-storey buildings having a fire resistant spandrel. *MATEC Web Conf.* **2016**, *46*, 04003. [CrossRef]

21. Barnett, C.R. Fire Separation between External Walls of Buildings. *Fire Saf. Sci.* **1989**, *2*, 841–850. [CrossRef]

22. Cicione, A.; Walls, R.; Sander, Z.; Flores, N.; Narayanan, V.; Stevens, S.; Rush, D. The Effect of Separation Distance between Informal Dwellings on Fire Spread Rates Based on Experimental Data and Analytical Equations. *Fire Technol.* **2020**. [CrossRef]

23. Modak, A.T. Thermal Radiation from Pool Fires. *Combust. Flame.* **1977**, *29*, 177–192. [CrossRef]

24. Hwang, C.H. Development of Water Curtain System Based on the Design of Optimized Fire Prevention Block in the Conventional Markets, Korea Fire Institute: Seoul, South Korea, 2016. Available online: https://www.ndsl.kr/ndsl/search/detail/report/reportSearchResultDetail.do?cn=TRKO201600011230 (accessed on 30 June 2016).

25. Cote, A.E.; Hall, J.R.; Powell, P.; Grant, C.C.; Solomon, R.E. *Fire Protection Handbook*, 20th ed.; National Fire Protection Association (NFPA): Quincy, MA, USA, 2008; Section 6; pp. 6–274. ISBN 978-087-765-758-3.

26. Wickström, U. The plate thermometer—A simple instrument for reaching harmonized fire resistance tests. *Fire Technol.* **1994**, *30*, 195–208. [CrossRef]

27. ISO 834-1:1999. Fire Resistance Tests—Elements of Building Construction—Part 1: General Requirements. 1999. Available online: https://www.iso.org/obp/ui/#iso:std:iso:834:-1:ed-1:v1:en (accessed on 3 March 2016).

28. Ingason, H.; Wickström, U. Measuring incident radiant heat flux using the plate thermometer. *Fire Saf. J.* **2007**, *42*, 161–166. [CrossRef]

29. Häggkvist, A. The Plate Thermometer as a Mean of Calculating Incident Heat Radiation. Master's Thesis, Luleå University of Technology, Luleå, Sweden, 2009.

30. Ditch, B. Evaluating pool fire severity and the cooling effect of local water spray using a continuous plate thermometer. *Fire Saf. Sci.* **2011**, *109*, 133–143. [CrossRef]

31. Han, H.S.; Yun, H.S.; Hwang, C.H. Calibration of the Plate Thermometer for measuring Heat Flux using a Conical Heater. *J. Mech. Sci. Technol.* **2019**, *33*, 3563–3569. [CrossRef]

32. NFPA 92B. Standard for Smoke Management Systems in Malls, Atria, and Large Spaces. National Fire Protection Association (NFPA). 2009. Available online: https://www.nfpa.org/codes-and-standards/all-codes-and-standards/list-of-codes-and-standards/detail?code=92B (accessed on 11 August 2015).
33. Poon, L.S. Predicting Time of Flashover. 1991. Available online: Iafss.org/publications/aofst/3/283/view/aofst_3-283.pdf (accessed on 2 July 2020).

Article

Numerical Investigation of the Turbulent Wake-Boundary Interaction in a Translational Cascade of Airfoils and Flat Plate

Xiaodong Ruan [1,*], Xu Zhang [1], Pengfei Wang [2], Jiaming Wang [1] and Zhongbin Xu [3]

[1] State Key Laboratory of Fluid Power and Mechatronic Systems, Zhejiang University, Hangzhou 310027, China; 3110104687@zju.edu.cn (X.Z.); wjm0@zju.edu.cn (J.W.)

[2] School of Engineering, Zhejiang University City College, Hangzhou 310015, China; wangpf@zucc.edu.cn

[3] Institute of Process Equipment, Zhejiang University, Hangzhou 310027, China; xuzhongbin@zju.edu.cn

* Correspondence: xdruan@zju.edu.cn; Tel.: +86-1385-805-3612

Received: 24 July 2020; Accepted: 27 August 2020; Published: 31 August 2020

Abstract: Rotor stator interaction (RSI) is an important phenomenon influencing performances in the pump, turbine, and compressor. In this paper, the correlation-based transition model is used to study the RSI phenomenon between a translational cascade of airfoils and a flat plat. A comparison was made between computational results and experimental results. The computational boundary layer velocity is in reasonable agreement with the experimental velocity. The thickness of boundary layer decreases as the RSI frequency increases and it increases as the fluid flows downstream. The spectral plots of velocity fluctuations at leading edge $x/c = 2$ under RSI partial flow condition $f = 20$ Hz and $f = 30$ Hz are dominated by a narrowband component. RSI frequency mainly affects the turbulence intensity in the freestream region. However, it has little influence on the turbulence intensity of boundary layer near the wall. A secondary vortex is induced by the wake–boundary layer interaction and it leads to the formation of a thickened laminar boundary layer. The negative-vorticity wake also facilitates the formation of a thickened boundary layer while the positive-vorticity wake has a similar effect, like a calmed region which makes the boundary layer thinner.

Keywords: rotor stator interaction; boundary layer; secondary vortex

1. Introduction

Rotor stator interaction (RSI) is a common phenomenon in pumps and turbomachines, especially in the case of multistage compressors. Many works have been carried out to investigate the influence of RSI on energy calculation [1], pressure fluctuation [2], internal fluid flow [3], noise [4], and machine vibration for turbomachines [5]. The unsteady interaction has an important influence on the hydrodynamic performance of turbomachines in many aspects such as blade loading, stage efficiency, heat transfer and noise generation [6]. On the one hand, the wake generated by the trailing edge of moving blades is periodically cut by the stationary blades, and pressure fluctuation is created, pressure fluctuation on the blade may lead to fatigue failure due to periodic load variation. On the other hand, the unsteady interaction changes the local pressure distribution, the inverse pressure gradient, and viscous effects may lead to the separation of the boundary layer where a back-flow region is generated, which makes the energy loss increase and the efficiency decrease. Therefore, in order to improve the efficiency and control the blade load, it makes sense to carry out an investigation of the mechanism by which RSI influences the boundary layer and how the boundary layer develops with the interaction.

Rotor stator interaction might be divided into two different interaction mechanisms, namely potential flow interaction and wake interaction. In general, wake interaction ranges farther downstream, but if the impellers and guide vanes are closely spaced, both mechanisms will occur simultaneously [7].

However, if the impellers and guide vanes are closely spaced, the fluid-induced force will be a very high level which will quickly lead to blade failure [8], so a close space design between rotor and stator is usually avoided. As a result, the primary mechanism of rotor stator interaction is wake interaction. Wakes generated by upstream impeller trailing edge convey low-speed fluids, and wakes are transported to the downstream where they interact with the boundary layer of the downstream guide vane in form of travelling wave [9]. This interaction will directly affect the development of the boundary layer of downstream blades. Although many investigations on boundary layer profiles have been made both theoretically and experimentally [10–12], the nature of the mechanism of wake–boundary interaction and its impact on the unsteadiness of boundary layer, including velocity profiles, turbulence intensity distribution, shear stress in different phases, and boundary layer structure, is not well understood.

In recent decades, there have been many theories on rotor stator interaction [13–15], but two procedures on rotor stator interaction are widely accepted. One was proposed by Kubota [16] and Franke [17] based on the diametrical mode theory. In this procedure, the periodic flow induced by the wake effects could be represented by the Fourier series, and rotor stator interaction can be calculated by the Fourier series superposition, whereby the diametrical mode number k could indicate the number of high or low-pressure fluid regions of the specific frequency component in the circumferential direction. This procedure can indicate the relative amplitude of each harmonic frequency components with different rotor/stator blade number [18]. The other procedure was proposed by Rodriguez [19] based on the sequence of interactions. This approach enables us to regard the vibration as the result of a modulation in the amplitudes of the interactions and rather than a consequence of an exciting diametrical mode number as the first theory proposed. Furthermore, this procedure could help to determine the origin of the harmonics, thus providing guidance for obtaining lower amplitude at specific harmonics or to avoid the generation of harmonics [20]. However, due to the complexity of the internal flow in turbomachines, the interaction is not only a consequence of potential-wake flow interaction but also a consequence of wake–boundary interaction. These two procedures neglect the unsteady wake–boundary interaction and naturally are both lack of engineering cases to corroborate its approach.

The combination of experiment and numerical calculation is an important method to study unsteady flow field [21–23]. To precisely compute the boundary layer flow, the choice of numerical methodology is very important. Sanders [24] made a comparison between computational fluid dynamics (CFD) predictions using Reynolds-averaged Navier–Stokes equations (RANS) and experimental results of unsteady wake effects on a highly loaded low-pressure turbine blade. And the results showed that the RANS method, especially the shear stress transport (SST) model, had a good prediction on the unsteady L1A blade flow field. Due to the shortcomings of the RANS method, the LES method has also recently been used to study the interaction of wake boundary layers if more accurate results are desired. Sarkar [25] used large eddy simulation (LES) to investigate the influence of wake structures on the boundary layer evolution at the suction side of a high-lift low-pressure turbine blade, and his research illustrated that in addition to the kinematics of wakes, the high- pressure oscillation and the curl of the separated shear layer along nearly half of the suction surface were determined by the length of the convective wake. Direct numerical simulation (DNS) is acknowledged to be the most perfect numerical method in accordance with experiments, because it is a direct solution to the N-S equation. However, due to its high spatial and temporal resolution, it is computationally intensive, time-consuming, and highly dependent on computer memory. Currently using DNS to simulation is still in the exploratory stage. Wissink [26] deliberated the effect of wake turbulence intensity on transition in a compressor cascade by using DNS, and the results indicated that the separation was intermittently inhibited due to the intense wake reaction as the periodically passing wakes tried to induce turbulent spots upstream of separation point. Particle Image Technology (PIV) has high measurement accuracy due to non-contact and small interference, so it has become a hot topic in recent fluid mechanics measurement research. Jia [27] conducted a PIV measurements of rotor boundary layer development with the effects of wakes generated by inlet guide vane (IGV), the research found that for the stronger wake, the interaction between the unsteady wake and the separated boundary layer on the suction surface of the airfoil was dominated by the high turbulence of wake and negative jet behavior of the wake. As the wake strength increased, the separation was almost totally inhibited. The different boundary layer states also affected the development of disturbances conversely.

From the above previous studies, we know that it is essential to study the wake generation and its effects on the boundary layer development. Before attempting the extremely complex three-dimensional and rotating flow on a turbomachine, the present study based on a simpler setup can be used as a preliminary orientation for the turbomachinery blade. This setup is abstracted from the blade row interaction problem of turbomachinery, the advantages are the geometrical complexity and pressure gradients are removed and it is convenient for calculations. The present study used the experimental data in the literature [28], which has convenient measurements and accurate data in the boundary layer. Comparisons were made between the experimental data and the CFD results to ensure the numerical simulation is reliable. The main objective of this work is to investigate the mechanism of the wake effects on boundary layer development, especially the RSI frequency effects on the boundary layer and its time–space variation.

2. Model and Methods

2.1. Computational Domain and Mesh Generation

The experimental model of turbomachine stage in the literature [28] was represented by a low speed wind tunnel with an unsteady wake generating rig. Although this model is a simple test rig, the model is quite representative of general realistic. When the turbomachinery impellers are expanded along the circumferential direction, the impeller movements can be considered as vertical movements, as shown in Figure 1. The stator blade was replaced by a flat plate in the test section, and the rotor blade was represented by a translational cascade of airfoils (NACA0024). The moving airfoils generate periodic wake disturbances and spread to downstream, so the wake reacts with the flat plate when it passes through the test section.

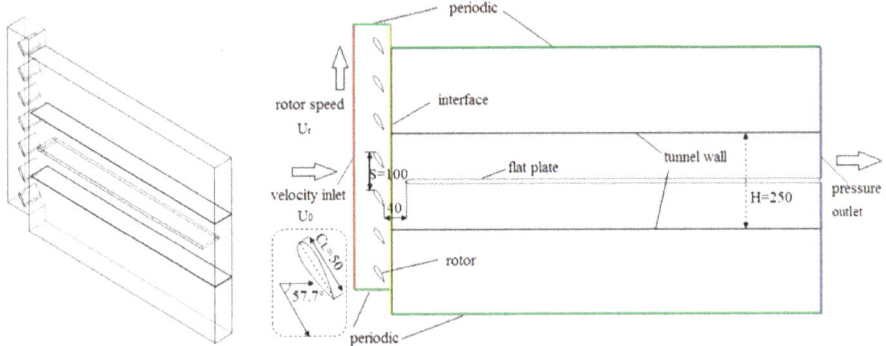

Figure 1. A global view of the computational domain, (**left**) 3D schematic of geometry (spanwise× 10), (**right**) plane view with boundary conditions.

The entire fluid domain consists of the moving cascade of airfoils and the stationary flat plate, as shown in Figure 1. 3D calculation was carried out in this study. In order to find a balance between the quality of simulation and computational cost, a geometric optimized 3D mesh with spanwise-reduced domain was set up to perform these numerical simulations. As a guide [29], the spanwise depth of the calculated domain was set as $LS = 0.16c$. Block-structured with O-grid around the airfoil and flat plate was adopted for the mesh designs. The γ-$Re_{\theta t}$ transition SST model was used in the transient numerical simulation. Therefore, the grid cells near the wall were encrypted with a cell stretching factor equal to 1.2 from the walls to ensure the value y+ less than 1 at the first node from the solid walls, as is required by the SST turbulence model [30,31]. The maximum number of layers is 25, and the wall y+ distribution is shown in Figure 2. It can be observed that the average value of y+ is 0.2 for airfoil and 0.4 for plate. A snapshot of mesh encryption is shown in Figure 3, it shows that the mesh of leading edge and trailing edge of airfoil and the boundary layer are encrypted. The quality of mesh is larger than 0.7 and 0.8 for airfoil domain and plate domain correspondingly, thus a good mesh quality is guaranteed. CFD simulations were performed with commercial software ANSYS Fluent 16.0 [32] which was run on an HP Z840 high-performance workstation using a total of 32 cores with 64GB memory for the unsteady calculations. These simulations required approximately 500 core-hours for all unsteady simulations on the Workstation with 3.1GHz Intel Xeon(R) CPU E5-2687w v3 (2 CPU).

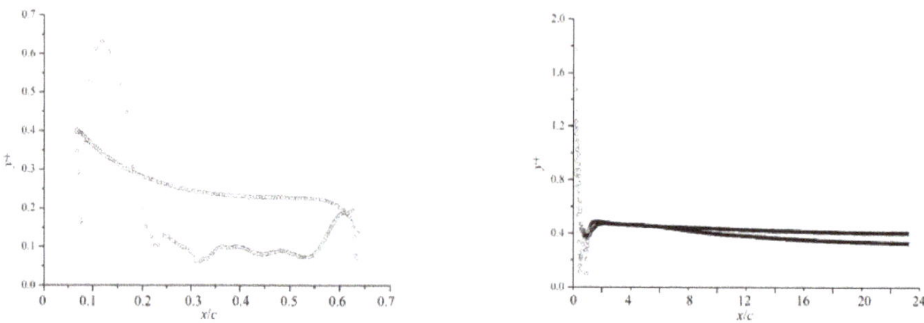

Figure 2. Wall $y+$ distribution of airfoil blade (**left**) and flat plate (**right**).

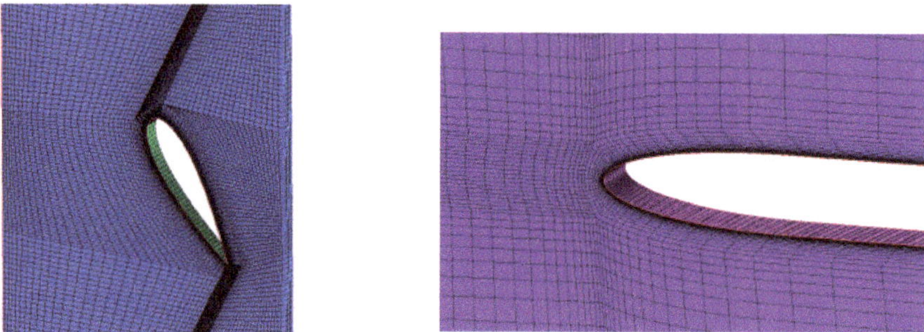

Figure 3. Snapshot of boundary layer mesh encryption (**left**: airfoil blade, **right**: flat plate).

2.2. Theoretical Methods

The viscous sublayer flow can be directly solved by the SST k-ω turbulence model, The SST k-ω model combined with turbulent shear stress transmission and is quite accurate in the prediction of flow separation near the wall under an inverse pressure gradient [33], however, this model has some defects in the prediction of the transition phenomenon [34]. Based on an empirical formula, the γ-$Re_{\theta t}$ transition SST model proposed by Langtry et al. [35] is more accurate in the prediction of the location of the transition point. This model takes advantage of the local parameters provided by a RANS solver rather than employing an additional boundary layer solver. The γ-$Re_{\theta t}$ transition SST model [36] is coupled with another two transport equations based on the SST k-ω model, and it does not influence the flow filed of fully turbulent region. So, theoretically, the transition SST model has higher prediction accuracy for the velocity distribution in the near-wall region. Many studies [37–39] have confirmed the accuracy of this model.

The transport equations of turbulent kinetic energy and specific dissipation rate are:

$$\frac{\partial(\rho k)}{\partial t} + \frac{\partial(\rho u_i k)}{\partial x_i} = P_k + D_k + \frac{\partial}{\partial x_i}\left[(\mu + \sigma_k \mu_t)\frac{\partial k}{\partial x_i}\right] \tag{1}$$

$$\frac{\partial(\rho \omega)}{\partial t} + \frac{\partial(\rho u_i \omega)}{\partial x_i} = P_\omega + D_\omega + \frac{\partial}{\partial x_i}\left[(\mu + \sigma_\omega \mu_t)\frac{\partial \omega}{\partial x_i}\right] \tag{2}$$

where P_k and P_ω are the production terms, D_k and D_ω are the destruction terms, u_i is the flow velocity, ρ is the density of fluid, μ is the dynamic viscosity of the fluid, σ_k and σ_ω are the model coefficients, k is the turbulent kinetic energy, and ω is the specific dissipation rate.

The γ-$Re_{\theta t}$ transition model can be written as:

$$\frac{\partial(\rho \gamma)}{\partial t} + \frac{\partial(\rho u_i \gamma)}{\partial x_i} = P_\gamma + E_\gamma + \frac{\partial}{\partial x_i}\left[\left(\mu + \frac{\mu_t}{\sigma_f}\right)\frac{\partial \gamma}{\partial x_i}\right] \tag{3}$$

$$\frac{\partial(\rho \tilde{Re}_{\theta t})}{\partial t} + \frac{\partial(\rho u_i \tilde{Re}_{\theta t})}{\partial x_i} = P_{\theta t} + \frac{\partial}{\partial x_i}\left[\sigma_{\theta t}(\mu + \mu_t)\frac{\partial \tilde{Re}_{\theta t}}{\partial x_i}\right] \tag{4}$$

where transport Equation (3) is about intermittency γ and Equation (4) is about the local transition onset momentum thickness Reynolds number $\tilde{Re}_{\theta t}$. P_γ and E_γ are the production term and the destruction term of the transport equation for intermittency γ, respectively. $P_{\theta t}$ is the source item of the transport equation for the local transition onset momentum thickness Reynolds Number $\tilde{Re}_{\theta t}$, μ is the dynamic viscosity coefficient, μ_t is the eddy viscosity, σ_f and $\sigma_{\theta t}$ are the constant coefficients.

2.3. Numerical Scheme

In order to get the accurate boundary layer velocity distribution, a reasonable boundary condition setting is very important. The computational domain is based on the parameters in the experiment. As the previous section states, the model is a representative of general realistic turbomachinery. The flow conditions were chosen based on the typical demand conditions. The "velocity inlet" was used in the tunnel inlet with velocity = 3 m/s according to the experiments. The turbulence parameters were specified according to the turbulence intensity and hydraulic diameter of the inlet. The "pressure outlet" was set in the outlet of the wind tunnel with pressure = 0 Pa, and the turbulence parameters were specified according to turbulence intensity and viscosity ratio. Besides, a smooth no-slip wall condition was specified for the solid wall boundaries of the airfoil surface, the flat plate wall, and the wind tunnel walls. The upper and lower boundaries were specified to be "periodic" boundary conditions. The front face and back face in the spanwise direction were set as symmetrical. The rationality of the periodic boundary is explained as follows: the rotor part itself is periodic, and the upper and lower areas of the stator side are fully expanded to match the rotor area. The extension of the stator area does not affect the stator flow field. The periodic boundary is added to the virtual extension to match the rotor area for calculation. The rotor-stator interface was set to periodic repeats; the detailed information of boundary condition settings is shown in Figure 1.

The RSI frequency f is defined as the ratio of the vertical velocity of moving airfoil U_r to the distance S between two adjacent airfoils. The undisturbed free-stream velocity, U_0, is set to 3 m/s for all simulation, and the vertical velocity of the rotor, U_r, ranged between 2 m/s and 4 m/s, so the RSI frequency f ranged from 20~40 Hz. The Reynolds number is 2.37×10^4 based on the half height of wind tunnel. Different rotor velocity corresponds to different incidence angles. The incidence angle is the largest at f = 20 Hz, so it is under partial flow condition, while the incidence angle at f = 40 Hz is the smallest, so in this case it is under full flow condition. The turbulence intensity in the wind tunnel is 0.7% in experiments, so a turbulence intensity of 0.7% is also applied at the velocity inlet during the calculation. The steady calculation is performed before the transient simulation was carried out, and the results of steady calculation are taken as the initial condition of the transient simulation.

The unsteady flow is calculated by a second-order upwind spatial discretization with a second-order implicit velocity formulation. And a pressure-based solver is applied, the SIMPLE pressure-velocity coupling method is used for the numerical simulations. The convergence criteria of 1×10^{-5} is used for the residual of continuity equations, momentum equations, and turbulence variables. The time step is set to 1×10^{-4} s so that there are 250, 333, and 500 time steps in a blade-passing cycle for $U_r = 4$ m/s, 3 m/s and 2 m/s correspondingly. To avoid the influence of unstable effects at the beginning, time-averaging process and analysis are made after the computation is conducted for at least 5 cycles. The velocity is recorded at points that are at a distance of $x/c = 2, 6, 10, 14$ from the leading edge of the flat plate.

3. Validation of the Numerical Model

Verification of mesh dependency was carried out to ensure the mesh has enough resolution. 3 different mesh resolutions were tested including coarse mesh case 3 (2.4×10^5 elements), standard mesh case 2 (3.6×10^5 elements) and fine mesh case 1 (4.8×10^5 elements), and the results are compared with experimental results as shown in Figure 4. Ultimately, the fine mesh case 1 was used for current computations. This proved that the numerical methodology used in this study is feasible.

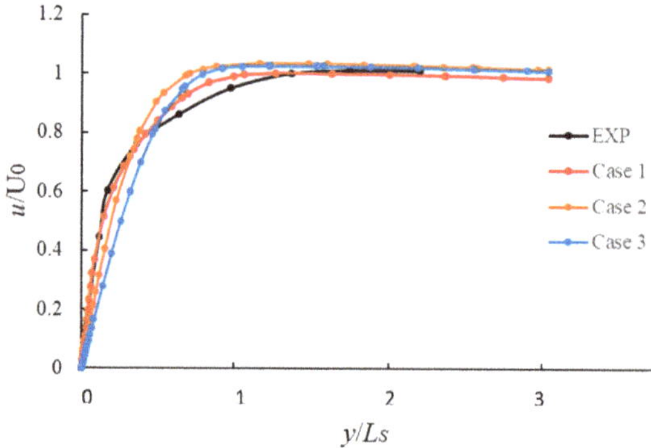

Figure 4. Boundary layer velocity profiles calculated by three different mesh sizes (steady condition).

In order to validate the accuracy of the numerical method, different turbulence models including Re- normalization group (RNG) k-ε turbulence model which has a quite high accuracy in predicting vortex flow, SST k-ω turbulence model which has the ability to predict the near wall flow, and transition SST model were chosen to predict the boundary layer velocity profiles at rotor-stator interaction frequency $f = 30$ Hz. The experimental velocity in the literature [28] was measured by a hot wire probe. U_0 is the velocity in the freestream, the thickness of boundary layer which is donated as δ_{99} is the distance from the wall of the plate to the region where $U = 0.99U_0$. For transient flow field calculations, the arithmetic mean of the fluid velocity at different time steps can be expressed as

$$U_{mean} = \frac{\sum_{i=1}^{N} u_i}{N} \tag{5}$$

where U_{mean} is time-averaged mean velocity, u_i is the transient velocity, and N is the calculated time steps.

The time-averaged predicting results in Figure 5a shows the mean velocity distribution in the boundary layer. Comparing the calculated results, the RNG k-ε turbulence model overestimates the velocity in the free stream flow by approximately 10%, the flow continuity was checked at each streamwise position, and the error was less than 1%. The reason why the RNG k-ε turbulence model overestimates the velocity in the free-flow region by 10% is that the RNG k-ε turbulence model underestimated the velocity in other flow regions. The results showed that the RNG k-ε turbulence model severely underestimated the velocity in the large velocity gradient region. Results calculated by the SST k-ω turbulence model and the transition SST model showed a more precise velocity prediction which has no more overestimation in the freestream region and less error within the large velocity gradient region. Figure 5b shows the boundary layer thickness distribution at different plate streamwise locations. The numerical results of the transition SST model coincide with the experimental results better, the boundary layer thickness ranges from 0.68 *y/Ls* to 1.31 *y/Ls* calculated by transition SST turbulence model at RSI frequency *f* = 30 Hz. As the fluid is transported downstream, the thickness of boundary layer increases gradually. Figure 6 is the time history contrast of velocity calculated by different turbulence model and from experimental results at *x/c* = 2, *f* = 20 Hz. All of the computational results can capture the pulsation in velocity. However, it is evident that the RNG k-ε turbulence model excessively overestimated the velocity. The SST k-ω turbulence model and the transition SST model show a similar prediction in the transient velocity. Ultimately, comparing the three predicting results with the experimental results, it can be concluded that the transition SST turbulence model is more accurate, and thus it is reasonable to use this turbulence model in this paper.

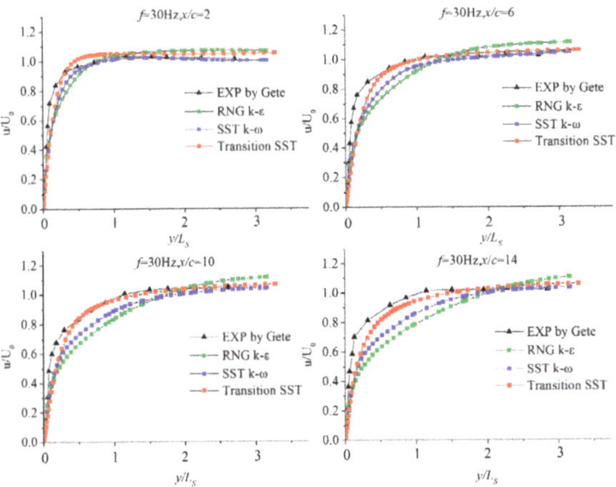

(**a**) Time-averaged mean velocity distribution of boundary layer at different streamwise positions.

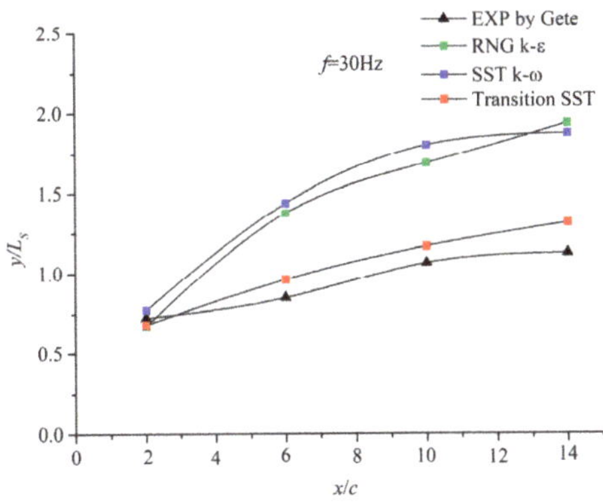

(**b**) Boundary layer thickness δ_{99} at different streamwise positions.

Figure 5. Comparison of boundary layer distribution between results calculated by different turbulence model and experimental results.

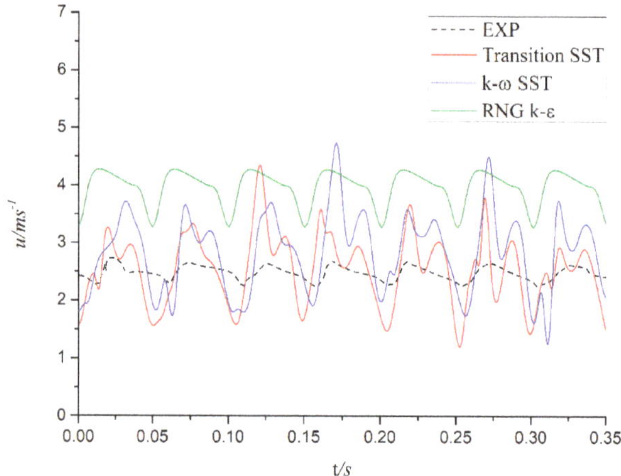

Figure 6. Time history of instantaneous velocity calculated by different turbulence model and experimental results at $x/c = 2$, $y/Ls = 0.25$, $f = 20$ Hz.

4. Results and Discussions

4.1. Velocity Profiles in the Boundary Layers

The statistical time-averaging process is made with results obtained from multiple periodic cycles. Figure 7a is the comparison of time-averaged mean velocity distributions calculated by different rotor-stator interaction frequency. It can be clearly observed that the velocity profiles of boundary layer are similar in the streamwise direction on the plate. Firstly, the velocity increases rapidly near the wall, and then the velocity growth rate decreases. When the wall distance y is greater than the thickness of the boundary, the velocity maintains the maximum and remains unchanged. For the three simulations with different RSI frequency, the velocity gradient in the boundary layer decreases from the leading edge of the plate at $x/c = 2$ to the trailing edge of plate at $x/c = 14$. This is mainly because the boundary layer of the flat plate has not been fully developed, the thickness of the boundary layer is still increasing as shown in Figure 7b, so its velocity gradient decreases with the increase of the streamwise distance. Moreover, the RSI frequency range investigated in this study shows that the significant fluctuation of time-average mean velocity distribution is a function of RSI frequency in unsteady flow. It can be observed that the velocity gradient in the boundary layer at $f = 40$ Hz is more significant than the velocity gradient at $f = 20$ Hz. This is mainly due to the decrease of incidence angle at $f = 40$ Hz, which results in a smaller scale wake vortex at the trailing edge of airfoil, and the thickness increasement of boundary layer caused by wake–boundary interaction decreases as shown in Figure 7b. The detailed boundary layer thickening mechanism will be discussed in later section. Figure 7b also shows the thickness of the laminar boundary layer and turbulent boundary layer of the flat plate according to Equations (6) and (7) [40]. It can be observed that the thickness of boundary layer is mostly between the thickness of laminar boundary layers and turbulent boundary layers due to the influence of the rotor wakes.

$$\delta_1(x) = 5\sqrt{\frac{vx}{U_\infty}} \tag{6}$$

$$\delta_2(x) = 0.37\left(\frac{v}{U_\infty}\right)^{1/5} x^{4/5} \tag{7}$$

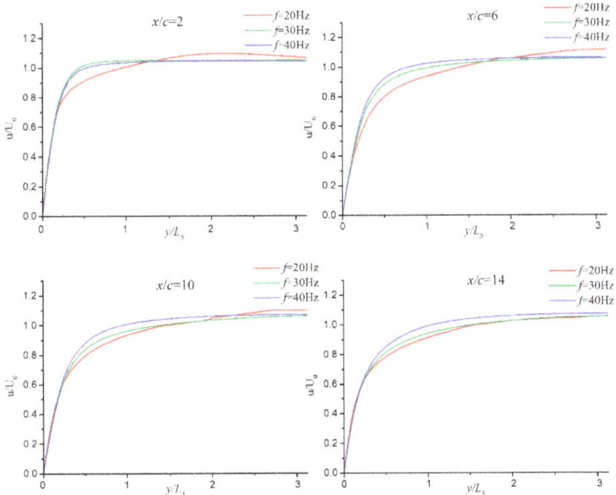

(**a**) Time-averaged mean velocity distribution of boundary layer for different RSI frequency at different streamwise position.

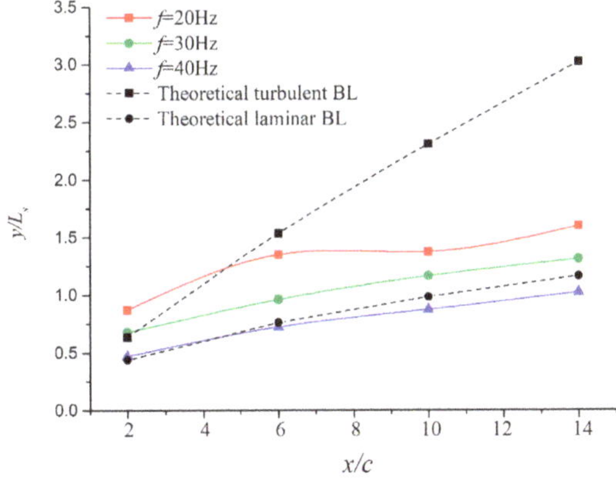

(**b**) Boundary layer thickness δ_{99} varies with the streamwise position for different RSI frequency.

Figure 7. Comparisons of boundary layer calculated by different rotor-stator interaction frequency.

Figure 8 is the transient velocity fluctuation in the streamwise direction and wall-normal direction. The spanwise fluctuations are not shown as the amplitude of spanwise fluctuations are far smaller than for the streamwise and wall-normal fluctuations. The reason why the spanwise fluctuation is so small is that the movement of airfoil blade causes the fluid to pulsate in the streamwise direction and wall-normal direction, and there is no source of disturbance in the spanwise direction except for turbulent random pulsation, which is far less than the periodic fluctuations. The transient results show that as the wake impinges on the boundary layer of flat plate, the velocity in the boundary layer shows obvious periodic fluctuations. The fluctuation amplitudes of u_x' and u_y' at the leading edge $x/c = 2$ are far greater than that at trailing edge $x/c = 14$. The amplitude of fluctuation velocity decreases with the increase of streamwise direction, this is mainly due to the fluctuation energy is viscously dissipated when the fluctuation spreads downstream. And it is evident that the fluctuation amplitude of u_x' is much higher than that of u_y'. For the flat plate, the velocity fluctuation amplitude in the streamwise direction and the wall-normal direction depends on two aspects. On the one hand, the installation angle of the front airfoil blade has an effect on the velocity fluctuation amplitude. The smaller the angle between the airfoil chord and the plate, the higher the fluctuation amplitude in wall-normal direction generated by the vertical movement of the airfoil blade. On the other hand, the velocity fluctuation depends on the inflow attack angle of the plate. The smaller the inflow attack angle, the lower the absolute amplitude of the velocity fluctuation in the wall-normal direction. Obviously, in the case of present study, it is mainly due to the small inflow attack angle of the plate, which causes the amplitude of the velocity fluctuation in wall-normal direction to be much lower than that of the streamwise direction. For the three RSI frequencies, the amplitude of fluctuation velocity at $f = 20$ Hz is the highest while the amplitude of fluctuation velocity at $f = 40$ Hz is the lowest. This is because the inflow attack angle of airfoil blade at $f = 20$ Hz is larger than that of $f = 40$ Hz, when the inflow attack angle of the airfoil blade is relatively large, the flow separation is liable to occur at the boundary layers, and large scale of vortex shedding is apt to form, which will lead to a high amplitude of velocity fluctuation.

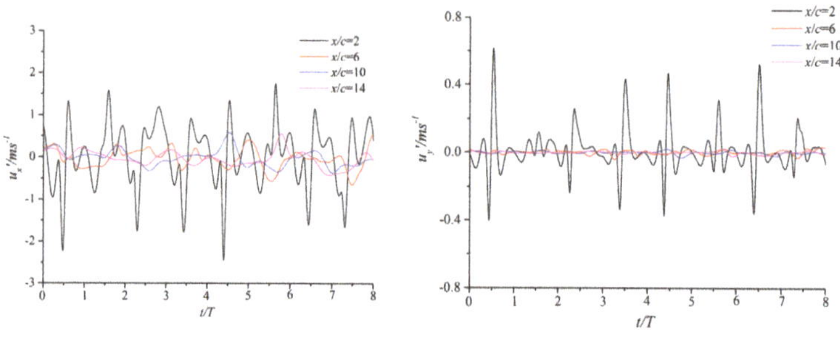

(**a**) Instantaneous velocity fluctuation at $f = 20$ Hz, left: u_x', right: u_y'.

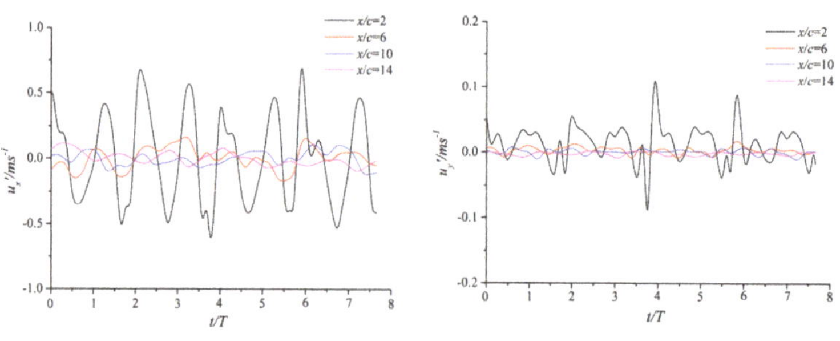

(**b**) Instantaneous velocity fluctuation at $f = 30$ Hz, left: u_x', right: u_y'.

Figure 8. *Cont.*

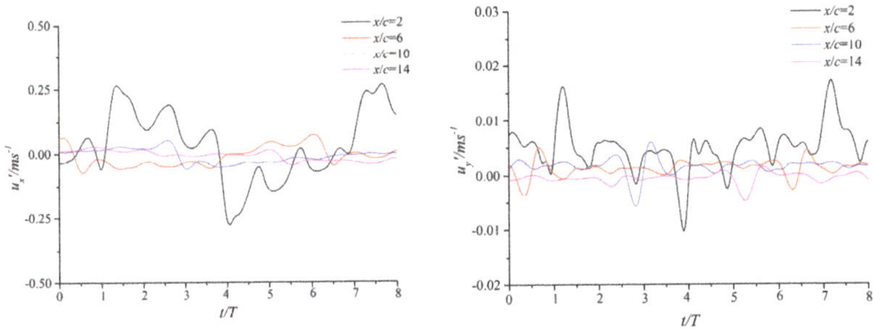

(**c**) Instantaneous velocity fluctuation at $f = 40$ Hz, left: u_x', right: u_y'.

Figure 8. Instantaneous velocity fluctuation at $y/Ls = 0.25$ over 8 passing cycle: (**a**) $f = 20$ Hz (**b**) $f = 30$ Hz (**c**) $f = 40$ Hz.

Due to the isotropic properties of turbulent flow in free flow, the fluctuation amplitude of u_x', u_y', u_z' should be approximately equal. However, because the specific motion of the rotor airfoil blades imposes different degrees of perturbation on the fluid in different directions, the wake conveys this perturbation into the boundary layer of the flat plate, which makes the fluctuation amplitude in the boundary layer of the flat plate vary greatly. Further perceptions into the growth of the fluctuations in the flow streamwise direction is revealed by spectral analysis. Figure 9 shows the spectral plots of velocity fluctuations at the leading edge $x/c = 2$ and the trailing edge $x/c = 14$. It can be observed that the fluctuation amplitudes of u_x', u_y' at $x/c = 2$ with RSI frequency $f = 20$ Hz are dominated by a narrowband component centered around 20 Hz. However, the fluctuation amplitude of u_z' does not exist a peak at a certain frequency. The structure of the fluctuation is dominated by an increasingly broadband spectrum. This is because the movement of rotor airfoil induces a relatively large disturbance of the fluid in the streamwise and wall-normal directions, but the disturbance in the spanwise direction is minimal. At the trailing edge $x/c = 14$, the frequency component with high peak amplitude disappears in the three directions, indicating that the high peak frequency component is dissipated. The structure of fluctuations becomes increasingly chaotic, resulting in the disappearance of narrowband component. The velocity fluctuation is dominated by an increasingly broadband spectrum. At $f = 30$ Hz, the spectral plots of velocity fluctuation are similar to the case of $f = 20$ Hz. When RSI frequency increases to $f = 40$ Hz, the velocity fluctuations at the leading edge $x/c = 2$ and trailing edge $x/c = 14$ are both dominated by an increasingly broadband spectrum. This is mainly induced by the decrease of the frequency component at $f = 40$ Hz, so there is no high-peak component in the entire fluctuation spectral plots. As the frequency of RSI increases, the difference of velocity fluctuation amplitude with u_x', u_y', and u_z' at the leading edge of the plate $x/c = 2$ also becomes larger. However, the amplitude of fluctuation at the trailing edge $x/c = 14$ has little to do with the variation of RSI frequency.

4.2. Turbulence Intensity in the Boundary Layer

Many experimental results have confirmed that the velocity of wake is quite low compared with the surrounding fluid, the wake is a negative jet with high turbulence [41]. When the convective wake with increased turbulence intensity permeates the laminar boundary layer on the plate surface, it disturbs the velocity in the boundary layer as the wake conveys high turbulence kinetic energy. The turbulence intensity in the plate boundary layer is investigated and the results are shown in Figure 10. Turbulence intensity profiles are similar for the three different frequencies. The turbulence intensity in the boundary layer increases first and then decreases. A peak value exists in the boundary layer. In the viscous sublayer, the viscosity stress is dominated for fluid motion. The viscous damping has a strong inhibitory effect on the tangential velocity fluctuation and normal velocity fluctuation.

As the distance from the wall increased, the turbulent Reynolds stress increases rapidly, so the turbulence intensity increases rapidly in the buffer layer. This indicates that the buffer layer is the primary turbulence energy producer; it provides the needed energy for the maintenance of wall turbulence. The outer region is fully developed turbulent flow. The turbulence intensity dissipated as the wall distance increase, and finally, it maintains a relatively low level in the outer region. The turbulence intensity profile is quite different from others for $f = 20$ Hz at $x/c = 2$. The profile is much fuller than others. At $f = 20$ Hz, the incidence angle of fluid around the airfoil is the largest, so flow separation occurs in the forward suction surface of airfoil, thus larger scale wake vortexes which contains high turbulence level are generated, wake vortex shedding impacts on the leading edge of the plate. The large scale of wake vortex makes the turbulence intensity profile much fuller. For the three RSI frequencies, the value of turbulence intensity is the highest at $f = 20$ Hz, nearly double the value at $f = 40$ Hz in the outside region. However, their peak value has a small difference. This indicates that the RSI frequency has little influence on the turbulence intensity in the near-wall boundary layer; the turbulence intensity near the wall might be affected by the wall shape. RSI frequency mainly affects the turbulence intensity in the freestream flow region.

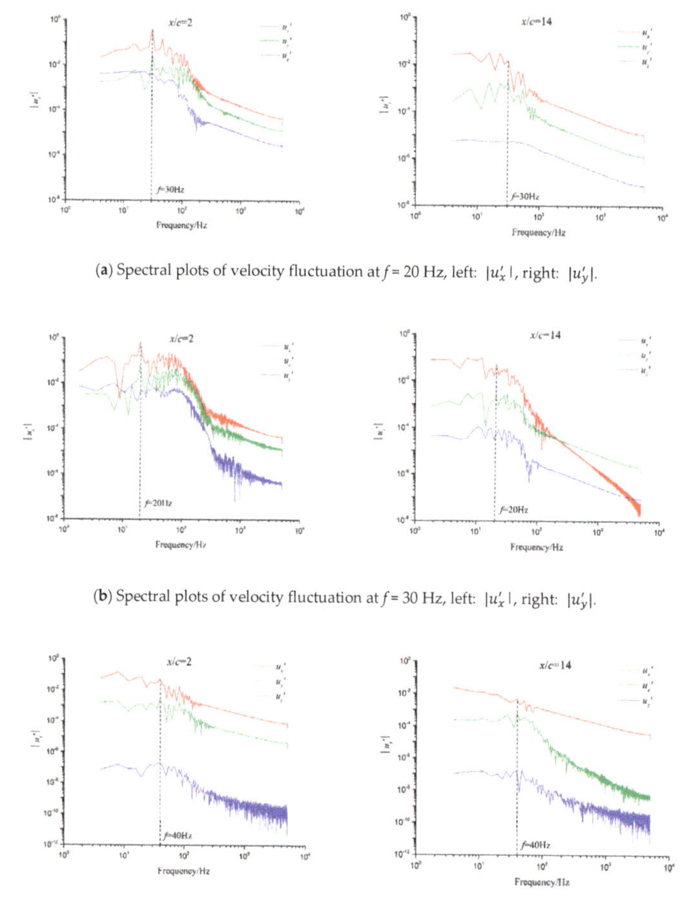

(**a**) Spectral plots of velocity fluctuation at $f = 20$ Hz, left: $|u'_x|$, right: $|u'_y|$.

(**b**) Spectral plots of velocity fluctuation at $f = 30$ Hz, left: $|u'_x|$, right: $|u'_y|$.

(**c**) Spectral plots of velocity fluctuation at $f = 40$ Hz, left: $|u'_x|$, right: $|u'_y|$.

Figure 9. Spectral plots of velocity fluctuation at $y/Ls = 0.25$ with different RSI frequency: (**a**) $f = 20$ Hz (**b**) $f = 30$ Hz (**c**) $f = 40$ Hz.

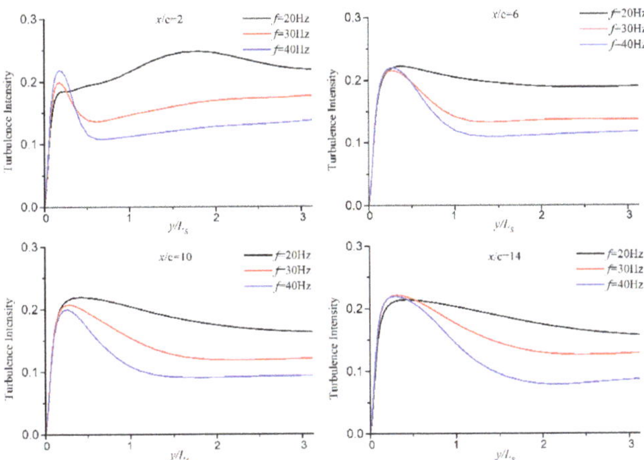

Figure 10. Turbulence intensity distribution in the boundary layer at different streamwise position.

Figure 11a–c is the time-space diagram of turbulence intensity calculated by different RSI frequency. The horizontal axis is the plate length; the ordinates are the time normalized by the passing period. Figure 11d is the schematic diagram of turbulence intensity at f = 40 Hz which outlines the boundary layer states. At this moment, it is necessary to introduce the boundary layer states described by Halstead et al. [42]. Due to the relative movement of blade rows, the wake from upstream airfoils periodically impacts the surface of the downstream plate, thus make the turbulence intensity in the plate periodically increase and decrease. The turbulence spots are initiated at some locations downstream from the leading edge by the wake disturbance. Region C in low turbulence intensity is a laminar region between wakes. Region A was used to represent the turbulence initiated by the wake. It represents a "wake-induced transitional strip". The boundary becomes turbulent when transition is completed within the region A, the strip continues to propagate to the trailing edge of the plate as a "wake-induced turbulent strip" labeled as E, in which turbulence intensity is decreasing along the streamwise direction. The turbulence spots are followed by a calmed region B. The high shear stress in calmed region B is considered to be one of the characteristics that can be effectively inhibit flow separation and delay transition. The region D followed the calmed region B represents the transition between wakes, and it finally becomes a turbulent region F. The calmed region extends to x/c = 5, 2.2, 2 from the leading edge for f = 20 Hz, f = 30 Hz, and f = 40 Hz correspondingly. Although the calmed region at f = 20 Hz is extended, it also could be found that in two adjacent periods, the area of the calmed region in one period is reduced a lot. Moreover, this result can be observed at f = 30 Hz as well. The calmed region in two adjacent periods is not equal until the RSI frequency increases to f = 40 Hz. This is mainly because the size of wake at f = 20 Hz and f = 30 Hz is quite large, so the trailing edge of the last wake on the boundary interacts with the leading edge of the next wake. As a result, it makes the laminar flow between wakes turns into turbulent flow in advance.

4.3. Convection of Wake and Wake-Induced Structure

By means of analyzing the distributions of the phase-averaged results, the whole process of turbulent wake–boundary layer interaction in a wake cycle can be precisely revealed. Figures 12–14 show the phase-averaged results of selected moments during one passing cycle with different RSI frequency. Three representative phases in a passing cycle are selected to show the wake movement of wakes and the interaction with the boundary layer. The first column is the velocity distribution in three selected phases, the second column is the turbulence kinetic energy distribution, and the third column is the z vorticity distribution.

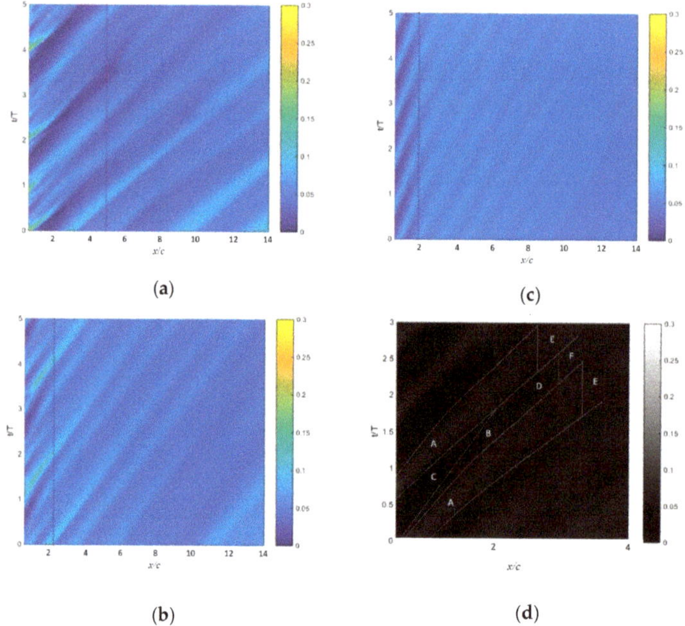

Figure 11. Time-Space diagram of turbulence intensity at $y/Ls = 0.25$ for different frequency (**a**) $f = 20$ Hz (**b**) $f = 30$ Hz (**c**) $f = 40$ Hz (**d**) schematic of boundary layer states at $f = 40$ Hz.

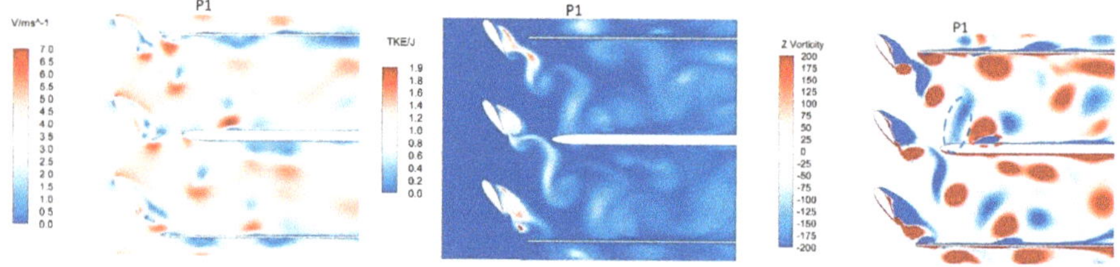

(**a**) Contours at phase 1, left: velocity, middle: turbulence kinetic energy, right: z vorticity.

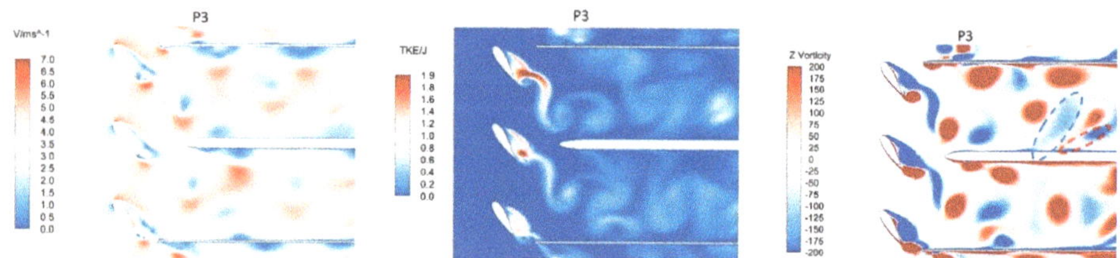

(**b**) Contours at phase 2, left: velocity, middle: turbulence kinetic energy, right: z vorticity.

Figure 12. *Cont.*

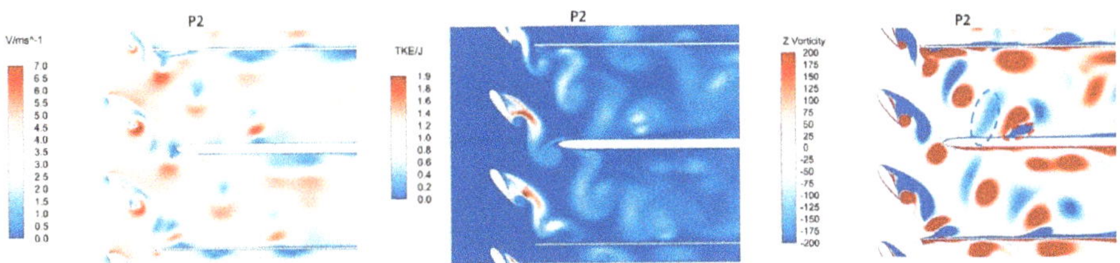

(**c**) Contours at phase 3, left: velocity, middle: turbulence kinetic energy, right: z vorticity.

Figure 12. Distributions of phase-averaged results at three selected phases at f = 20 Hz.

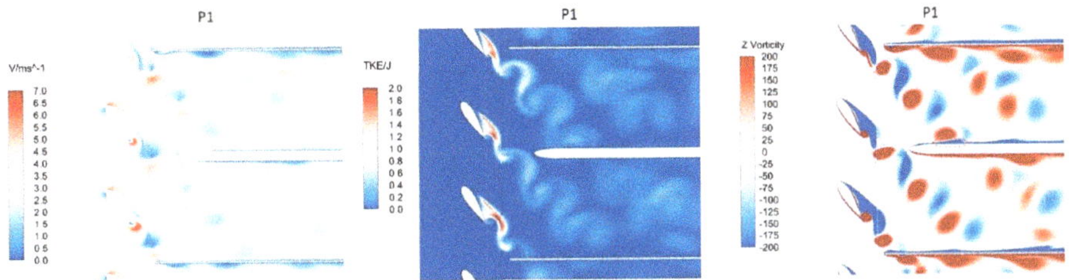

(**a**) Contours at phase 1, left: velocity, middle: turbulence kinetic energy, right: z vorticity.

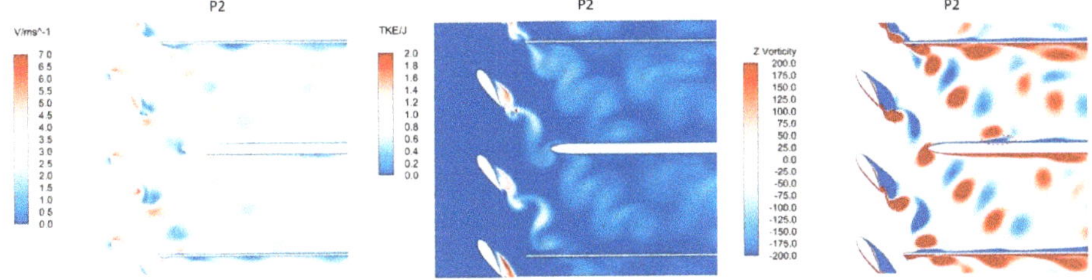

(**b**) Contours at phase 2, left: velocity, middle: turbulence kinetic energy, right: z vorticity.

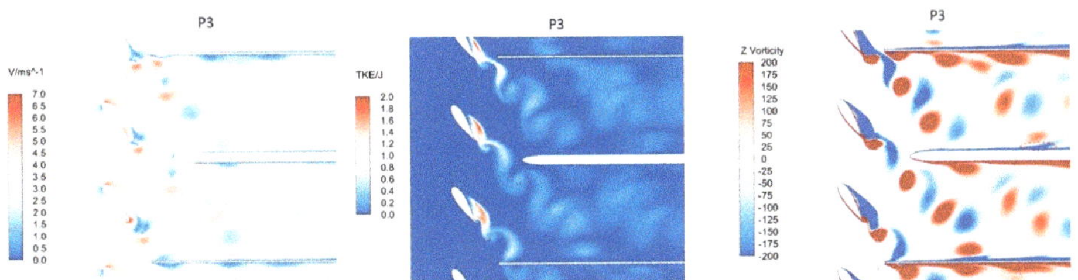

(**c**) Contours at phase 3, left: velocity, middle: turbulence kinetic energy, right: z vorticity.

Figure 13. Distributions of phase-averaged results at three selected phases at f = 30 Hz.

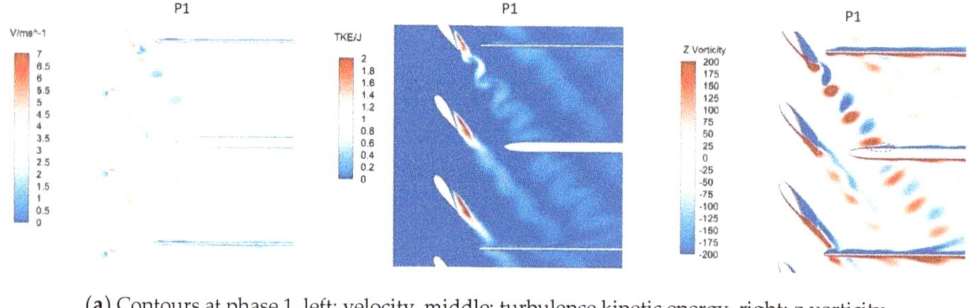

(**a**) Contours at phase 1, left: velocity, middle: turbulence kinetic energy, right: z vorticity.

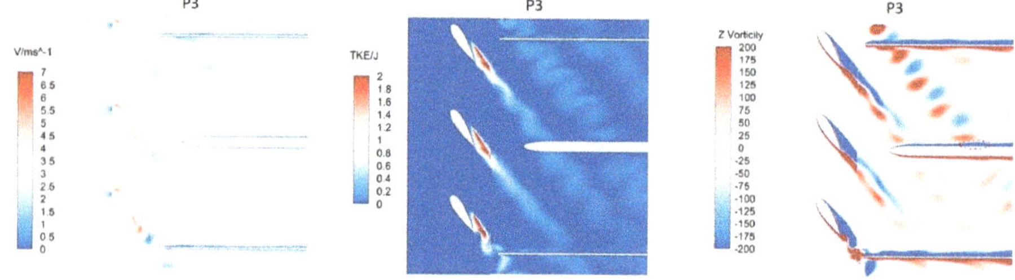

(**b**) Contours at phase 2, left: velocity, middle: turbulence kinetic energy, right: z vorticity.

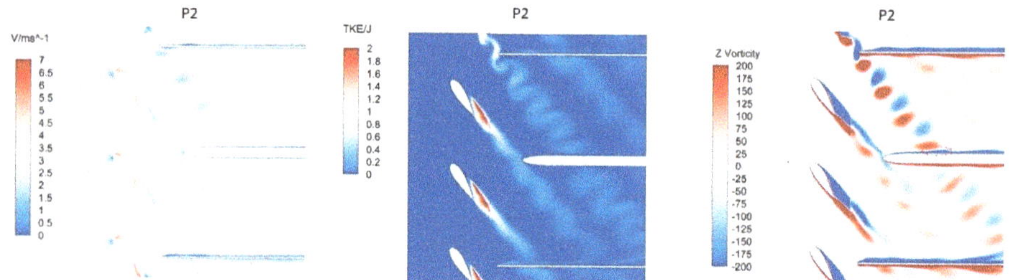

(**c**) Contours at phase 3, left: velocity, middle: turbulence kinetic energy, right: z vorticity.

Figure 14. Contours of phase-averaged results at three selected phases at f = 40 Hz.

As many experimental results have shown [43], the velocity distributions show that the wake is a negative jet with high turbulence kinetic energy. Moreover, it can be observed from the vorticity distribution results in Figures 12–14 that the wake is opposite around the centerline, the vorticity from the pressure surface is positive while the vorticity from the suction surface is negative. The vorticity distributions on both sides of the airfoil are asymmetrical due to the different flow on the pressure surface and suction surface at the trailing edge of airfoil. Comparing the phase-averaged results in different RSI frequency, it is found that the rotor wake is twisted and produces a pocket vortex at the end of wake vortex and a thickened boundary layer at the front of wake vortex. The thickened boundary layer has more concentrated vorticity and higher turbulence kinetic energy than the phase-averaged results of the wake. From the velocity distribution in Figures 12–14, it is observed that there are a high-speed region and a low-speed region in the surrounding area of the thickened boundary layer, and those flow structures move downstream along the plate boundary together.

When the wake vortex is close enough to the wall, the pressure gradient is strong enough to make the local fluid separate, thus a bubble containing vorticity of inverse sense to main vortex is produced (Figure 12a). Propagating downstream, this bubble grows rapidly to the separation point where it departs from the wall and evolves into a secondary vortex (red dotted line), which is supplied by a vortex sheet from the separation point (Figure 12b). However, the secondary vortex at f = 30 Hz and f = 40 Hz has been always attached to the wall and has no separation. This is mainly because the wake vortex strength at f = 30 Hz and f = 40 Hz is weakened compared with that of f = 20 Hz. According to the theory of Doligalski [44], the fractional convection rate α increases as the absolute strength of the vortex is decreased. For α < 0.75, in the region behind the vortex, a rapid lifting of the boundary-layer streamlines and strong growth of boundary-layer occurs. The fractional convection rate α < 0.75 at f = 20 Hz while α > 0.75 at f = 30 Hz and f = 40 Hz. So, the secondary vortex induced by wake vortex keep attached to the plate surface. What's more, the phase-averaged results could also explain the length difference of the calmed region shown in Figure 11. The calmed region of f = 20 Hz is obviously longer than that of f = 30 Hz and f = 40 Hz. This is because the tail of positive vorticity wake (wake between red dotted line and blue dotted line) interacts with the boundary layer like a calmed region, so the calmed region is extended. It decreases the thickness of the boundary layer. It can be observed that the tail of positive vorticity wake interacts with the boundary layer for a long distance at f = 20 Hz. In contrast, the positive vorticity wake interacts with the boundary layer for a pretty short distance at f = 30 Hz and f = 40 Hz. So, the calmed region at f = 30 Hz and f = 40 Hz is shorter compared with that of f = 40 Hz.

Figure 15 shows the boundary layer thickness δ_{99} at the leading edge x/c = 2 with different RSI frequency. The black dashed-dotted line represents the center of the wake, labeled as C. While the black dashed lines around the center line represent the leading and trailing edges of the wake, labeled as LE and TE. It is apparent that the thickness of boundary layer rapidly increases because of the presence of secondary vortex, the maximum thickness of boundary layer lies on the center line of secondary vortex. Because of the presence of calmed region, the thickness of boundary layer becomes thinner. However, due to the interaction of the negative vorticity vortex (Figure 13 blue dotted line), a low speed region is created on the plate surface, thus a platform appears on the decreasing curve. But there is no platform on the decreasing curve at f = 40 Hz, this is mainly because the negative vorticity vortex in the wake decreases in size and does not contact the plate surface. Finally, the thickness of boundary layer continues to decrease to a laminar boundary layer in the absence of wake interaction. It can be concluded that the wake itself cannot suppress separation and transition. However, the calmed region induced by the interaction between the airfoil wake and the boundary layer has the effects of suppressing separation and making the boundary layer thinner.

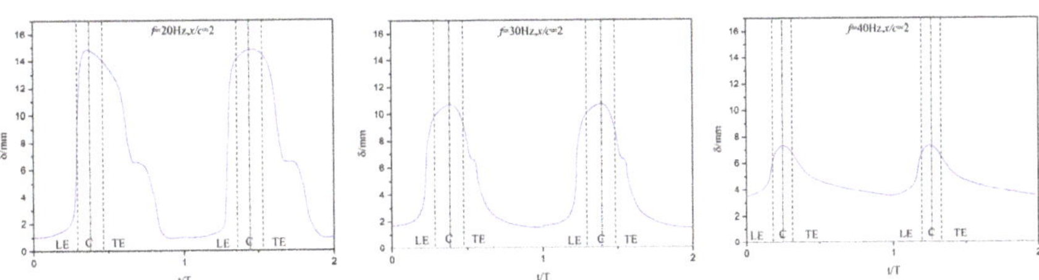

Figure 15. Boundary layer thickness δ_{99} variation in two periods at different RSI frequency.

5. Conclusions

The wake–boundary layer interactions in the turbomachine field are key fundamental issues from a physics perspective, i.e., the boundary layer behavior and properties with the effects of rotor wakes, as well as the engineering design, such as flow separation near the boundary layer and boundary layer heat transfer. The research in this article mainly focus on the fundamental physical mechanism of wake–boundary layer interaction. In order to extensively study the wake–boundary layer interaction, the γ-$Re_{\theta t}$ transition SST model is adopted in the transient numerical simulation of the 3D airfoil cascade and flat plate. Furthermore, the effects of different RSI frequency are investigated. The time-averaged results and transient results are compared with the experimental results, and the comparisons show that the numerical results have a reasonable agreement. This research deepens the understanding of the complex flow of wake–boundary layer interaction in turbomachinery. The results obtained in this article could provide some useful references for the design of turbomachinery. The main conclusions are as follows:

The boundary layer profiles are much fuller at RSI frequency f = 40 Hz. The boundary layer thickness of the flat plate decreases as the RSI frequency increases. The boundary layer thickness increases as the fluid flows downstream. As the wake impinges on the boundary layer of plate, the streamwise velocity fluctuations in the boundary layer are far greater than for the wall-normal fluctuation and spanwise fluctuation. For the three RSI frequency, the fluctuation amplitude at partial flow condition f = 20 Hz is the highest, while the fluctuation amplitude at full flow condition f = 40 Hz is the lowest. The spectral plots of $u_x{}'$, $u_y{}'$ at leading edge x/c = 2 under partial flow condition f = 20 Hz and f = 30 Hz are dominated by a narrowband component. However, the amplitude of fluctuation at the trailing edge x/c = 14 has little to do with the variation of RSI frequency and it is dominated by an increasingly broadband spectrum.

The turbulence intensity in the boundary layer increases first and then decreases; a peak value exists in the buffer layer. For the three RSI frequencies, the value of turbulence intensity is the highest at partial flow condition f = 20 Hz, nearly double the value at design condition f = 40 Hz in the outside freestream region. However, their peak value has a small difference. This indicates that the RSI frequency has little influence on the turbulence intensity in the near-wall boundary layer. The turbulence intensity might be affected by the wall shape and freestream velocity. RSI frequency mainly affects the turbulence intensity in the freestream region.

A secondary vortex with high turbulence kinetic energy is produced as the rotor wake interacts with the boundary layer of plate. And a calmed region is formed behind the secondary vortex. The calmed region has an effective suppression of the boundary layer transition. The calmed region at partial flow condition f = 20 Hz is far longer than that for design condition f = 40 Hz. This is mainly because the positive vorticity wake has a similar effect as a calmed region. The tail of the positive vorticity wake interacts with the boundary layer for a long distance at partial flow condition f = 20 Hz while it only continues a short distance at design condition f = 40 Hz. In the leading edge, the thickness of boundary layer rapidly increases because of the presence of the secondary vortex. The maximum thickness of boundary layer lies on the center of secondary vortex. Then the thickness of boundary layer decreases because of the calmed region. Next, the thickness of boundary layer goes into a platform as the negative-vorticity wake interacts with the boundary layer. Finally, the boundary layer becomes a laminar boundary layer in the absence of wake interaction.

Author Contributions: Conceptualization, X.R.; methodology, X.Z.; software, X.Z.; validation, J.W., Z.X.; formal analysis, X.Z.; investigation, X.R.; resources, X.R.; data curation, Z.X.; writing—original draft preparation, X.Z.; writing—review and editing, P.W.; visualization, X.Z.; supervision, X.R.; project administration, X.R.; funding acquisition, X.R. All authors have read and agreed to the published version of the manuscript.

Funding: This research was funded by the Science Fund for Creative Research Groups of National Natural Science Foundation of China [Grant No. 51821093] and the National Basic Research Program of China [Grant No. 2015CB057301].

Acknowledgments: We also highly appreciate reviewers for useful comments and editors for improving the manuscript.

Conflicts of Interest: The authors declare no conflict of interest.

References

1. Courtiade, N.; Ottavy, X.; Gourdain, N. Modal decomposition for the analysis of the rotor-stator interactions in multistage compressors. *J. Therm. Sci.* **2012**, *21*, 276–285. [CrossRef]
2. Vogeler, K.; Mailach, R. Rotor-Stator Interactions in a Four-Stage Low-Speed Axial Compressor-Part I: Unsteady Profile Pressures and the Effect of Clocking. *J. Turbomach.* **2004**, *126*, 833–845.
3. Gourdain, N. Prediction of the Unsteady Turbulent Flow in an Axial Compressor Stage. Part 2: Analysis of Unsteady RANS and LES Data. *Comput. Fluids* **2015**, *106*, 67–78. [CrossRef]

4. Hetherington, R. Compressor Noise Generated by Fluctuating Lift Resulting from Rotor-Stator Interaction. *AIAA J.* **2012**, *1*, 473–474. [CrossRef]

5. Dickmann, H.-P.; Wimmel, T.S.; Szwedowicz, J.; Filsinger, D.; Roduner, C.H. Unsteady Flow in a Turbocharger Centrifugal Compressor: Three-Dimensional Computational Fluid Dynamics Simulation and Numerical and Experimental Analysis of Impeller Blade Vibration. *J. Turbomach.* **2006**, *128*, 455–465. [CrossRef]

6. Hodson, H.P. Boundary Layer and Loss Measurements on the Rotor of an Axial-Flow Turbine. *J. Eng. Gas Turbine Power* **1984**, *106*, 391–399. [CrossRef]

7. Arndt, N.N.; Acosta, A.J.; Brennen, C.E.; Caughey, T.K. Rotor–Stator Interaction in a Diffuser Pump. *J. Turbomach.* **1989**, *111*, 213–221. [CrossRef]

8. Robinson, C.; Casey, M.; Hutchinson, B.; Steed, R. Impeller-Diffuser Interaction in Centrifugal Compressors. In Proceedings of the ASME Turbo Expo, Copenhagen, Denmark, 11–15 June 2012; Volume 8. [CrossRef]

9. Brear, M.J.; Hodson, H.P. The Effect of Wake Passing on a Flow Separation in a Low-Pressure Turbine Cascade. *J. Fluids Eng.* **2004**, *126*, 250–256. [CrossRef]

10. Czarske, J.R.; Ttner, L.B.; Razik, T.; Ller, H.M. Boundary layer velocity measurements by a laser Doppler profile sensor with micrometre spatial resolution. *Meas. Sci. Technol.* **2002**, *13*, 1979–1989. [CrossRef]

11. Oweis, G.F.; Winkel, E.S.; Cutbrith, J.M.; Ceccio, S.L.; Perlin, M.; Dowling, D.R. The mean velocity profile of a smooth-flat-plate turbulent boundary layer at high Reynolds number. *J. Fluid Mech.* **2010**, *665*, 357–381. [CrossRef]

12. Wheeler, A.P.S.; Miller, R.J.; Hodson, H.P. The Effect of Wake Induced Structures on Compressor Boundary-Layers. *J. Turbomach.* **2006**, *129*, 705–712. [CrossRef]

13. Gerolymos, G.A.; Michon, G.J.; Neubauer, J. Analysis and Application of Chorochronic Periodicity in Turbomachinery Rotor/Stator Interaction Computations. *J. Propuls. Power* **2002**, *18*, 1139–1152. [CrossRef]

14. Chaluvadi, V.S.P.; Kalfas, A.I.; Banieghbal, M.R.; Hodson, H.P.; Denton, J.D. Blade-Row Interaction in a High-Pressure Turbine. *J. Propuls. Power* **2001**, *17*, 892–901. [CrossRef]

15. Sato, W.; Yamagata, A.; Hattori, H. A Study on Unsteady Aerodynamic Excitation Forces on Radial Turbine Blade due to Rotor-Stator Interaction. In Proceedings of the 11th International Conference on Turbochargers and Turbocharging, London, UK, 13–14 May 2014; pp. 389–398.

16. Kubota, Y.; Suzuki, T.; Tomita, H.; Nagafugi, T.; Okamura, C. Vibration of Rotating Bladed Disc Excited by Stationary Distributed Forces. *Bull. JSME* **1983**, *26*, 1952–1957. [CrossRef]

17. Franke, G.; Fisher, R.; Powell, C.; Seidel, U.; Koutnik, J. On Pressure Mode Shapes Arising from Rotor-Stator Interactions. *Sound Vib.* **2005**, *39*, 14–18.

18. Li, D.; Gong, R.; Wang, H.; Wei, X.; Liu, Z.S.; Qin, D.Q. Analysis of Rotor-Stator Interaction in Turbine Mode of a Pump-Turbine Model. *J. Appl. Fluid Mech.* **2016**, *9*, 2259–2268. [CrossRef]

19. Rodriguez, C.; Egusquiza, E.; Santos, I.F. Frequencies in the Vibration Induced by the Rotor Stator Interaction in a Centrifugal Pump Turbine. *J. Fluids Eng.* **2007**, *129*, 1428–1435. [CrossRef]

20. Trivedi, C.; Cervantes, M.J. Fluid-structure interactions in Francis turbines: A perspective review. *Renew. Sustain. Energy Rev.* **2017**, *68*, 87–101. [CrossRef]

21. Keck, H.; Sick, M. Thirty years of numerical flow simulation in hydraulic turbomachines. *Acta Mech.* **2008**, *201*, 211–229. [CrossRef]

22. Li, C.B.; Jiang, X.Y. Flow Structures in Transitional and Turbulent Boundary Layers. *Phys. Fluids* **2019**, *31*, 111301. [CrossRef]

23. Ortolan, A.; Courty-Audren, S.K.; Binder, N.; Carbonneau, X.; Rosa, N.G.; Challas, F. Experimental and Numerical Flow Analysis of Low-speed Fans at Highly Loaded Wind Milling Conditions. *J. Turbomach.* **2017**, *139*, 071009. [CrossRef]

24. Sanders, D.D.; Nessler, C.A.; Sondergaard, R.; Polanka, M.D.; Marks, C.; Wolff, M.; O'Brien, W.F. A CFD and Experimental Investigation of Unsteady Wake Effects on a Highly Loaded Low Pressure Turbine Blade at Low Reynolds Number. In Proceedings of the ASME Turbo Expo 2010: Power for Land, Sea, and Air, Glasgow, UK, 14–18 June 2010.

25. Sarkar, S. Influence of Wake Structure on Unsteady Flow in a Low Pressure Turbine Blade Passage. *J. Turbomach.* **2009**, *131*, 041016. [CrossRef]

26. Wissink, J.; Zaki, T.A.; Rodi, W.; Durbin, P.A. The Effect of wake Turbulence Intensity on Transition in a Compressor Cascade. *Flow Turbul. Combust.* **2014**, *93*, 555–576. [CrossRef]

27. Jia, L.; Zou, T.; Zhu, Y.; Lee, C. Rotor boundary layer development with inlet guide vane (IGV) wake impingement. *Phys. Fluids* **2018**, *30*, 040911. [CrossRef]

28. Gete, Z.; Evans, R. An experimental investigation of unsteady turbulent-wake/boundary-layer interaction. *J. Fluids Struct.* **2003**, *17*, 43–55. [CrossRef]

29. Solís-Gallego, I.; Fernández, A.M.; Fernández-Oro, J.; Díaz, K.M.A.; Velarde-Suárez, S. LES-based numerical prediction of the trailing edge noise in a small wind turbine airfoil at different angles of attack. *Renew. Energy* **2018**, *120*, 241–254. [CrossRef]

30. Fernandez-Gamiz, U.; Mármol, M.G.; Rebollo, T.C. Computational Modeling of Gurney Flaps and Microtabs by POD Method. *Energies* **2018**, *11*, 2091. [CrossRef]

31. Thompson, J.F.; Warsi, Z.U.A.; Mastin, C.W. *Numerical Grid Generation*; Elsevier Science Publishing: London, UK, 1985.

32. Fluent Inc. *User's Guide*; Fluent 16.0; Fluent Inc.: New York, NY, USA, 2016.

33. Menter, F.R. Review of the shear-stress transport turbulence model experience from an industrial perspective. *Int. J. Comput. Fluid Dyn.* **2009**, *23*, 305–316. [CrossRef]

34. Zeng, Y.; Yao, Z.; Gao, J.; Hong, Y.; Wang, F.; Zhang, F. Numerical Investigation of Added Mass and Hydrodynamic Damping on a Blunt Trailing Edge Hydrofoil. *J. Fluids Eng.* **2019**, *141*, 081108. [CrossRef]

35. Langtry, R.B.; Menter, F.R. Correlation-Based Transition Modeling for Unstructured Parallelized Computational Fluid Dynamics Codes. *AIAA J.* **2009**, *47*, 2894–2906. [CrossRef]

36. Langtry, R.B.; Menter, F.R.; Likki, S.R.; Suzen, Y.B.; Huang, P.G.; Völker, S. A Correlation-Based Transition Model Using Local Variables—Part II: Test Cases and Industrial Applications. *J. Turbomach.* **2004**, *128*, 423–434. [CrossRef]

37. Gorji, S.; Seddighi, M.; Ariyaratne, C.; Vardy, A.E.; O'Donoghue, T.; Pokrajac, D.; He, S. A comparative study of turbulence models in a transient channel flow. *Comput. Fluids* **2014**, *89*, 111–123. [CrossRef]

38. Khayatzadeh, P.; Nadarajah, S. Laminar-turbulent flow simulation for wind turbine profiles using the γ–Reθt transition model. *Wind Energy* **2014**, *17*, 901–918. [CrossRef]

39. Furst, J.; Straka, P.; Příhoda, J.; Šimurda, D. Comparison of several models of the laminar/turbulent transition. *EPJ Web Conf.* **2013**, *45*, 1–6. [CrossRef]

40. Schlichting, H. *Boundary Layer Theory*, 7th ed.; McGraw-Hill: Boston, MA, USA, 1979.

41. Meyer, R.X. The Effect of Wakes on the Transient Pressure and Velocity Distributions in Turbomachines. *J. Basic Eng.* **1958**, *4*, 1544–1552.

42. Halstead, D.E.; Wisler, D.C.; Okiishi, T.H.; Walker, G.J.; Hodson, H.P.; Shin, H.-W. Boundary Layer Development in Axial Compressors and Turbines: Part 1 of 4—Composite Picture. *J. Turbomach.* **1997**, *119*, 114–127. [CrossRef]

43. Lu, X.; Zhang, Y.; Li, W.; Hu, S.; Zhu, J. Effects of periodic wakes on boundary layer development on an ultra-high-lift low pressure turbine airfoil. *Proc. Inst. Mech. Eng. Part A J. Power Energy* **2016**, *231*, 25–38. [CrossRef]

44. Doligalski, T.L.; Walker, J.D.A. The boundary layer induced by a convected two-dimensional vortex. *J. Fluid Mech.* **1984**, *139*, 1–28. [CrossRef]

Article

Unsteady Numerical Calculation of Oblique Submerged Jet

Weixuan Jiao [1], Di Zhang [1], Chuan Wang [1,*], Li Cheng [1,*] and Tao Wang [2]

[1] College of Hydraulic Science and Engineering, Yangzhou University, Yangzhou 225009, China; DX120170049@yzu.edu.cn (W.J.); DX120180054@yzu.edu.cn (D.Z.)

[2] Key Laboratory of Fluid and Power Machinery, Ministry of Education, School of Energy & Power Engineering, Xihua University, Chengdu 610039, China; mailtowangtao@163.com

* Correspondence: wangchuan198710@126.com (C.W.); chengli@yzu.edu.cn (L.C.)

Received: 31 July 2020; Accepted: 8 September 2020; Published: 11 September 2020

Abstract: A water jet is a kind of high-speed dynamic fluid with high energy, which is widely used in the engineering field. In order to analyze the characteristics of the flow field and the change of law of the bottom impact pressure of the oblique submerged impinging jet at different times, its unsteady characteristics at different Reynolds numbers were studied by using the Wray–Agarwal (W-A) turbulence model. It can be seen from the results that in the process of jet movement, the pressure at the peak of velocity on the axis was the smallest, and the velocity, flow angle, and pressure distribution remain unchanged after a certain time. In the free jet region, the velocity, flow angle, and pressure remained unchanged. In the impingement region, the velocity and flow angle decreased rapidly, while the pressure increased rapidly. The maximum pressure coefficient of the impingement plate changed with time and was affected by the Reynolds number, but the distribution trend remained the same. In this paper, the characteristics of the flow field and the law of the impact pressure changing with time are described.

Keywords: unsteady flow; numerical simulation; submerged jet

---

1. Introduction

The main advantages of an impinging jet are that it can effectively strengthen local heat and mass transfer, and thus has been widely used in many engineering fields [1–3] such as water conservancy and hydropower discharge engineering, in which the plunging nappe impacts the plunge pool, the cooling of electronic components, the chemical mixing jet, the micro-sprinkler irrigation in water-saving irrigation, etc. The impinging jet depends on the shape of the nozzle, the jet outlet speed, the impinging angle, the impinging height, and other parameters. Previous studies have shown that the flow field of the impinging jet mainly includes a free jet zone, an impingement zone, and a wall jet zone [4,5]. Moreover, with the change of parameters, the flow field in different regions presents different laws. Compared with a vertical impinging jet, the flow field and pressure distribution of the oblique impinging jet in the wall jet region are not uniform and are more complex [6,7].

With the improvement of computer performance, computational fluid dynamics technology provides a low-cost and high-efficiency solution to the complex problems in water conservancy engineering [8–10], municipal engineering [11,12], and other practical projects. Many scholars use Computational Fluid Dynamics (CFD) technology to study the flow field of an impinging water jet. Ghiti et al. [13] studied the jet impingement on a horizontal plate at different Reynolds numbers. The results showed that there is a positive correlation between vorticity and the Reynolds number. The turbulent characteristics near the wall plate are greatly affected by the Reynolds number. Based on the large eddy simulation (LES) turbulence technique, So et al. [14] carried out three-dimensional analysis of the flow mechanism of the unconfined plane impinging jet and provided a detailed flow structure. Gopalakrishnan et al. [15] studied round jet impingement at multiple impingement angles. The results showed that the larger the jet angle, the higher the jet growth that was predicted. Baghel et al. [16] presented the hydrodynamic structure and heat transfer performance of a vertical free surface jet impinging on a horizontal plate. By comparing the simulation results of the free surface jet velocity field with different turbulence models, it was found that the v^2f turbulence model can predict the flow field better than other turbulence models. There is also much research on the impact pressure of a water impinging jet. Wei et al. [17] studied the dynamic pressure caused by a submerged inclined jet by forced ventilation at the bottom of a deep plunge pool in detail, and proposed that the combination of a reduced jet thickness and an enhanced jet aeration can effectively reduce the impact on the bottom of a deep plunge pool. Tian et al. [18] investigated the impact pressure characteristics of jet equipment that can reach a maximum jet velocity of 50 m/s. Through numerical simulation, the reason for a scale effect of the impact pressure was elucidated. Shao et al. [19] improved the traditional smoothed particle hydrodynamics (SPH) method and verified its effectiveness by simulating the water jet impingement problem. The results showed that the improved SPH method can accurately simulate the shape and pressure distribution of an impinging jet. Based on the W-A turbulence model, Wang et al. [20] used CFD technology to study the flow characteristics of a submerged impinging water jet at different impact heights. The results show that there is a negative correlation between impact pressure, impact pressure coefficient and impact height. Through CFD technology, scholars can also carry out research on the unsteady flow process of water jets. Based on the incompressible Reynolds-averaged Navier–Stokes (RANS) equations with the W-A turbulence model, Zhang et al. [21] carried out numerical simulation of single-jet impingement. The results showed that the W-A model has good applicability in numerical simulation of single-jet impingement. Zhang et al. [22] studied the time-dependent initial flow structure of a subsonic unsteady elliptic jet using the large eddy simulation (LES) method. Based on the unsteady FLUENT(ANSYS Inc., Pittsburgh, PA, USA) numerical simulation technology, Yang et al. [23] carried out numerical simulation on the internal flow characteristics of different impinging jets. The results showed that the local Nusselt number oscillates over time. Chung et al. [24] studied the momentum and heat transfer characteristics of an unsteady impinging jet by using the numerical simulation method. The results showed that the impact heat transfer is very unstable due to the influence of the primary vortex generated by the jet nozzle. Zhang et al. [25] carried out numerical simulation of the internal flow field of the self-excited pulsed cavitating jet nozzle, analyzed the evolution process of the initial generation, energy concentration, and release of the cavitation in the chamber in one cycle, and explained the occurrence mechanism of the cavitating jet. Zaafori et al. [26] used the finite difference method to analyze the influence of the initial perturbation of the co-current flow on the jet dynamics and thermal behavior over time.

An oblique submerged jet is a complex flow with a practical engineering application background and important theoretical research value. The systematic study of it not only helps to deepen the understanding of the interaction mechanism between jet and wall, but also provides a scientific basis for engineering applications. In order to reveal the evolution law of the hydraulic characteristics of an oblique submerged jet flow field with time and the influence of the Reynolds number on its flow characteristics, the UDF (User Define Function) of FLUENT software was used to introduce the W-A model to simulate the unsteady flow field and pressure distribution of an oblique submerged impinging jet under different Reynolds numbers in this paper. In this study, a general circular jet, i.e., a fully developed circular jet, was selected for unsteady calculation.

2. Numerical Simulation

2.1. Geometric Model and Boundary Conditions

A three-dimensional non-closed oblique submerged impinging jet model was built, as shown in Figure 1. The jet was obliquely sprayed into a regular sink through a circular nozzle. The bottom surface was an impingement plate, the top surface was a free surface, the left and right sides were outlet boundaries, and the front and back sides were fixed wall boundaries. The inner diameter of the nozzle was $D = 20$ mm, and the outer diameter was $d = 22$ mm. All lengths in this paper were dimensionless with D; in order to ensure that the jet outlet flow was a fully developed turbulent pipe flow, the nozzle length was $50\,D$; the calculation domain of the sink was a length of $L_w = 55\,D$, a width of $W_w = 7\,D$, and a height of $H_w = 12\,D$; the vertical distance between the center of the nozzle outlet and the impingement plate was $H = 3\,D$. The calculation time step was 10^{-4} s, and the total simulation time was 1 s.

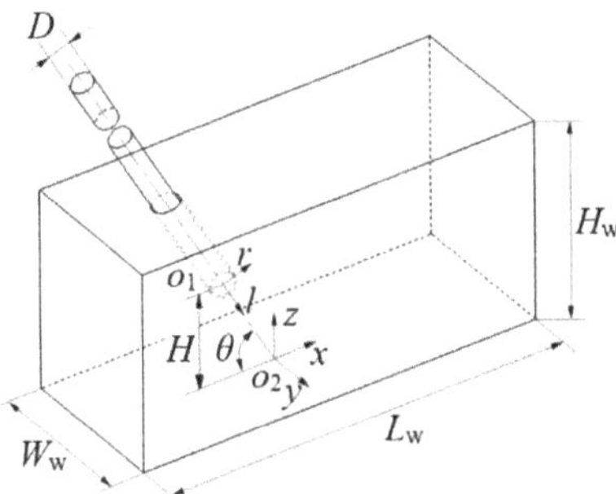

Figure 1. 3-D oblique submerged jet.

In order to better analyze the flow field of the oblique submerged impinging jet, as shown in Figure 1, the oblique submerged impinging jet was described in a double coordinate system. The rectangular coordinate system ro_1l was used to describe the impinging jet flow in the normal central plane before the impingement. The rectangular coordinate system o_2xyz was used to describe the impinging jet flow in the normal central plane after the impingement. The origin of ro_1l was set at the center of the nozzle outlet; r and l represent the radial and axial (along the jet) directions, respectively. The o_2xyz coincides with the geometric center (GC) of the jet on the impingement plane, with x, y, and z representing the longitudinal, transverse, and vertical directions, respectively.

The grid quality has a crucial influence on the numerical calculation. In this paper, ICEM software was used to divide the calculation domain into hexahedral grids. A grid sensitivity analysis was carried out to assess the required grid density. As shown in Table 1, several cases with different mesh cell numbers were compared. The minimum number of mesh cells was 2.65×10^6, and the maximum number of mesh cells was 4.80×10^6.

Table 1. Grid information of the different cases.

Case	1	2	3	4	5
Number of mesh cells in the calculation domain	2.65×10^6	3.21×10^6	3.73×10^6	4.26×10^6	4.80×10^6

Figure 1 shows the comparison between the CFD and the experimental results of the axial velocity V/V_b in the centerline of the jet. V_b denotes the bulk velocity at the jet exit, which can be defined as:

$$V_b = 4Q/\pi D^2 \tag{1}$$

where V represents the velocity at any position of the jet exit; Q is the flow rate of the jet.

As shown in Figure 2, when the number of mesh cells of the overall calculation domain reached 3.73 million, the difference between the CFD numerical simulation results and the experimental results was very small. After grid sensitivity analysis, the total number of mesh cells in the calculation domain was finally determined to be 4.26 million, as shown in Figure 3. The drawn grid can effectively reduce the false diffusion caused by poor grid quality, and can also improve the computing speed.

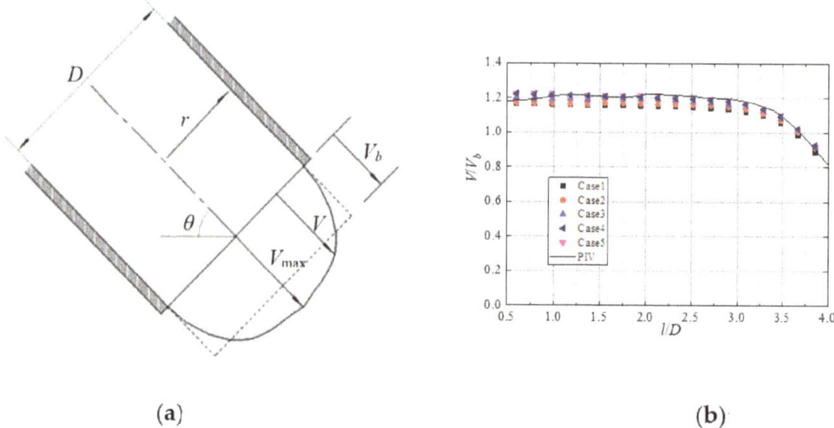

(**a**) (**b**)

Figure 2. (**a**) Definition of the bulk mean velocity (V_b); (**b**) grid independence test.

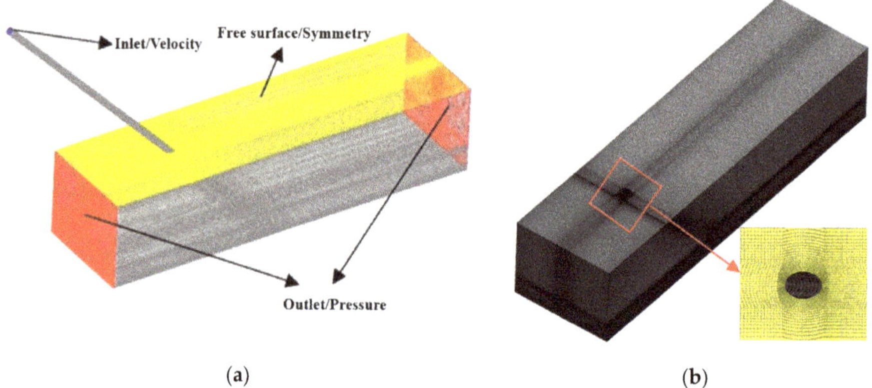

(**a**) (**b**)

Figure 3. Grid of the oblique submerged impinging jet: (**a**) Computational domain; (**b**) regular sink.

In the calculation, based on the pressure solver, the SIMPLE algorithm for pressure–velocity coupling was adopted. The pressure equation was discretized by the second-order scheme, and the momentum and turbulence model equations were discretized by the second-order upwind scheme. The turbulent viscosity ratio was used to define the turbulence on the boundary of the flow field. The inlet turbulent viscosity ratio was 10. The inlet adopted velocity inlet, the speed was evenly distributed, and the inlet working condition is shown in Table 2. The outlet was set as a pressure outlet, and the gauge pressure value was 0 Pa. This paper simulated the three-dimensional impinging jet in an open channel water environment, and its free surface remained almost unchanged, so the rigid-lid (RL) assumption method was used to process the free surface, and it was set as a symmetrical surface. A fixed wall with no slip was selected for the wall surface.

Table 2. Inlet boundary conditions.

	Condition 1	Condition 2	Condition 3
Inlet velocity (m/s)	0.585	1.17	1.76
Re	11,700	23,400	35,100

2.2. Wray–Agarwal Turbulence Model

2.2.1. Introduction of Wray–Agarwal Turbulence Model

The Wray–Agarwal turbulence model is a low Reynolds number one-equation model developed using the k-ω model. In this paper, the maximum y+ was less than 1 to ensure that the near wall treatment for the W-A model was accurate. Wang et al. [20] and Han et al. [27] carried out a comparison between numerical simulations and experimented on various turbulence models, including the W-A model. The results showed that the W-A model simulates the separation characteristics of the boundary layer more accurately and has good applicability in a numerical simulation of a submerged jet. The W-A model includes the cross-diffusion in the ω equation, and $R = \kappa/\omega$ is determined by the following transport equation:

$$\frac{\partial R}{\partial t} + \frac{\partial u_j R}{\partial x_j} = \frac{\partial}{\partial x_j}\left[(\sigma_R R + v)\frac{\partial R}{\partial x_j}\right] + C_1 R S + f_1 C_{2k\omega}\frac{R}{S}\frac{\partial R}{\partial x_j}\frac{\partial S}{\partial x_j} - (1 - f_1)C_{2k\varepsilon}R^2\left(\frac{\frac{\partial S}{\partial x_j}\frac{\partial S}{\partial x_j}}{S^2}\right) \tag{2}$$

The turbulent eddy viscosity:

$$\mu = \rho f_u R \tag{3}$$

where ρ is the density. S takes on the usual definition for mean strain:

$$S = \sqrt{2S_{ij}S_{ij}}\ S_{ij} = \frac{1}{2}\left(\frac{\partial u_i}{\partial x_j} + \frac{\partial u_j}{\partial x_i}\right) \tag{4}$$

Wall blocking is accounted for by the damping function:

$$f_\mu = \frac{\chi^3}{\chi^3 + C_W^3}, \tag{5}$$

where $\chi = R/v$ and $v = \mu/\rho$. The switching function is:

$$f_1 = min\left(tanh(arg_1^4), 0.9\right) arg_1 = \frac{1 + \frac{d\sqrt{RS}}{v}}{1 + \left[\frac{max\left(d\sqrt{RS}, 1.5R\right)}{20v}\right]^2} \tag{6}$$

where d is the minimum distance to the nearest wall.

The constants are:

$$\begin{aligned}
C_{1kw} &= 0.0829\ C_{1k\varepsilon} = 0.1127 \\
C_1 &= f_1(C_{1k\omega} - C_{1k\varepsilon}) + C_{1k\varepsilon} \\
\sigma_{k\omega} &= 0.72\ \sigma_{k\varepsilon} = 1.0 \\
\sigma_R &= f_1(\sigma_{k\omega} - \sigma_{k\varepsilon}) + \sigma_{k\varepsilon} \\
C_{2k\omega} &= \frac{C_{1k\omega}}{\kappa^2} + \sigma_{k\omega}\ C_{2k\varepsilon} = \frac{C_{1k\varepsilon}}{\kappa^2} + \sigma_{k\varepsilon} \\
\kappa &= 0.41\ C_\omega = 8.54
\end{aligned} \tag{7}$$

2.2.2. Verification of the Wray–Agarwal Turbulence Model

The Wray–Agarwal, Standard k-ε, RNG k-ε (Renormalization Group k-ε), Realizable k-ε, Standard k-ω, and SST k-ω (Shear Stress Transfer k-ω) turbulence models were used to simulate the oblique submerged impinging jet with an impact angle of $\theta = 45°$ and an impact height of H/D = 3. The numerical results were compared with the experimental data of PIV [28].

Figure 4 shows a comparison between the numerical velocity V/V_{max} with the empirical formula for a fully developed circular jet. The calculation formula is as follows:

$$V/V_{max} = (1-2r/D)^{1/n} \tag{8}$$

where V_{max} represents the maximum velocity at the jet exit, r represents the radial direction, and n is the empirical constant for the power law.

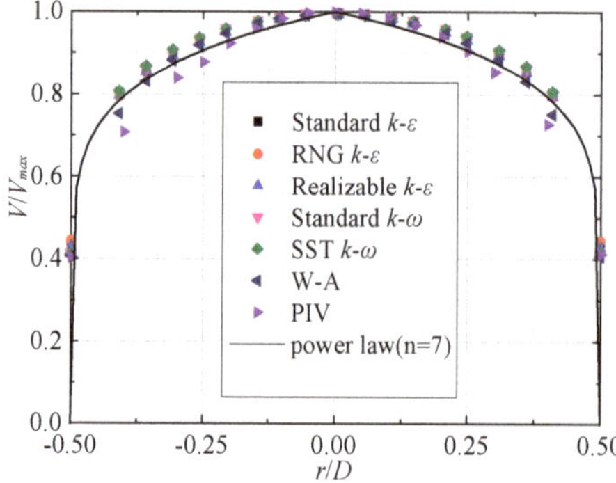

Figure 4. Comparison of calculation results with different turbulence models: Profiles of the normalized axial velocity V/V_{max} near the jet exit (l/D = 0.5).

When the value of n was 7, the velocity distribution calculated by Formula (8) was approximately consistent with the velocity distribution of a fully developed circular jet [29]. As shown in Figure 4, the numerical calculation results with different turbulence models are in good agreement with the empirical formula. This also shows that the flow at the jet exit was a fully developed jet flow. At the same time, the calculated results of the W-A model are closest to the empirical formula.

Table 3 illustrates the percentage difference between the CFD and the experimental results of the axial velocity V/V_b in the centerline of the jet. The validity of the turbulence model was verified by comparing the numerical simulation and experimental results of the axial velocity V/V_b in the centerline of the jet. As can be seen from Table 3, the calculated results of the W-A model are closest to the experimental results. Hence, the W-A model can accurately simulate the unsteady characteristics of an oblique submerged impinging jet.

Table 3. Numerical and experimental results with different turbulent models.

Turbulence Model	Standard k-ε	RNG k-ε	Realizable k-ε	Standard k-ω	SST k-ω	W-A
$l = 0.98\,D$	3.02%	1.91%	2.03%	1.85%	2.81%	0.80%
$l = 1.75\,D$	3.40%	2.28%	2.37%	2.22%	3.17%	0.16%
$l = 2.52\,D$	4.71%	3.54%	3.60%	3.51%	4.44%	1.46%
$l = 3.28\,D$	4.19%	2.68%	2.74%	2.92%	3.63%	1.23%

3. Results

3.1. Analysis of the Oblique Submerged Jet Flow Field under Different Times

Figure 5 shows the V/V_b velocity contours and streamline ($Re = 35{,}100$) of the mid-section ($y/D = 0$) of the jet at different times (0.01~1 s). When $t = 0.01$ s, the jet presented a circular arc shape distribution; when $t = 0.02$ s, the jet fluid moved axially, there was a vortex around the front of the jet, and the vortex had good symmetry with respect to the central axis; with the flow of the jet, the core position of the vortex gradually approached the impingement plate. When $t = 0.1$ s, most of the jets were deflected in the forward flow direction, the vortex in the forward flow direction gradually increased, and the vortex in the backward flow direction gradually decreased. With the increase of time t, the jet flowed along the wall, the vortex core remained near the front of the wall jet region in the forward flow direction, and the vortex gradually increased.

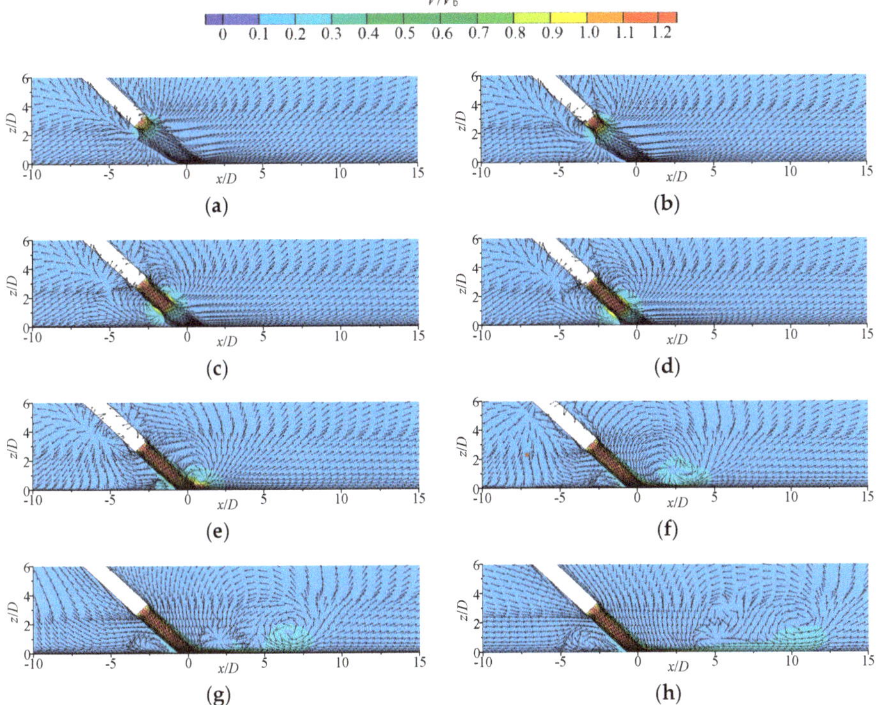

Figure 5. Normalized mean velocity V/V_b contour and streamline of the mid-section of the oblique submerged impinging jet at different times: (**a**) $t = 0.01$ s, (**b**) $t = 0.02$ s, (**c**) $t = 0.05$ s, (**d**) $t = 0.06$ s, (**e**) $t = 0.1$ s, (**f**) $t = 0.2$ s, (**g**) $t = 0.5$ s, (**h**) $t = 1.0$ s.

Figure 6 shows the vorticity contours of the mid-section of the oblique submerged impinging jet at different times ($Re = 35{,}100$). It can be seen that in the stage where the jet ejected the nozzle until it hit the impingement plate ($0 \leq t \leq 0.5$ s), the vorticity intensity was small. When the jet fluid collided with the impingement plate, the vorticity intensity increased in the backward flow direction. When comparing with Figure 4, it can be observed that when the jet deflected and flowed along the wall jet region, the vortex mainly occurred in the tongue part before the jet, and the vortex intensity was large. With the increase of time, the flow of the tongue part before the jet became complicated, and the vorticity distribution was irregular.

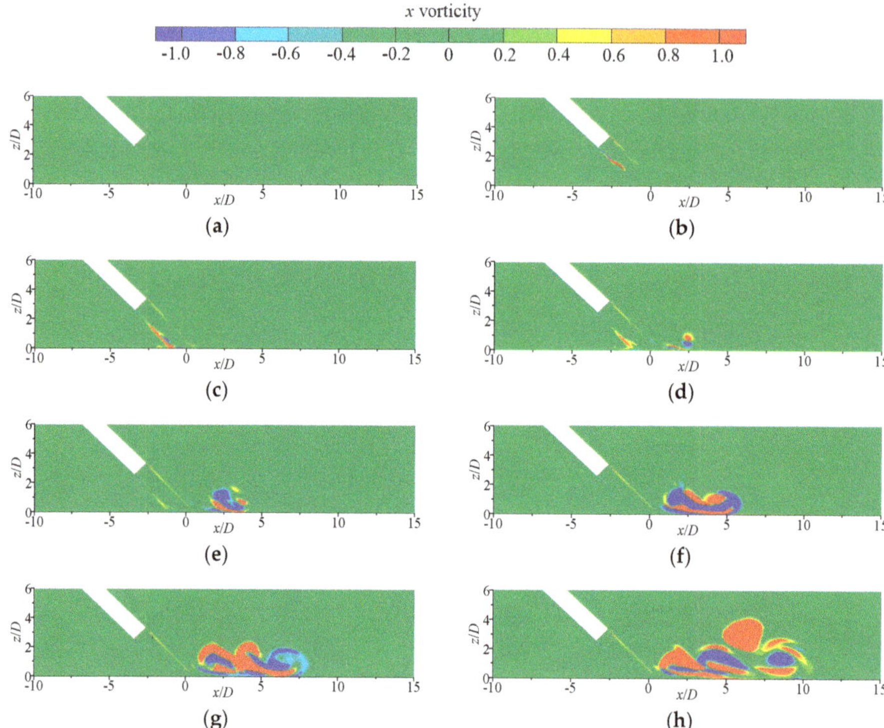

Figure 6. Vorticity of mid-section of the oblique submerged impinging jet at different times: (**a**) $t = 0.01$ s, (**b**) $t = 0.02$ s, (**c**) $t = 0.05$ s, (**d**) $t = 0.06$ s, (**e**) $t = 0.1$ s, (**f**) $t = 0.2$ s, (**g**) $t = 0.5$ s, (**h**) $t = 1.0$ s.

3.2. Comparison of the Oblique Submerged Jet Flow Field under Different Reynolds Numbers

Figures 7 and 8 show the dimensionless velocity V/V_b and flow angle φ distribution on the axis of the submerged jet at different times under different Reynolds numbers. As shown in Figure 7, under the condition of $Re = 11,700$, when 0 s $\leq t \leq 0.05$ s, the value of V/V_b remained unchanged, then decreased rapidly, and the influence range of jet increased; when $t = 0.1$ s, the jet velocity first remained constant, then rose, and finally dropped rapidly; when $t \geq 0.5$ s, the distribution of the V/V_b value was as follows: In the region of $0 \leq l/D \leq 3$, it remained unchanged, and in the region of $l/D \geq 3$, it decreased rapidly. When Re was increased, that is, when the speed of the jet nozzle was increased, the time for V/V_b to peak and the time for reaching the state where the V/V_b value did not change became shorter. It can be seen from Figure 8 that with the increase of time, the flow angle in the free jet region gradually increased to the impinging angle and then remained unchanged, and the decrease rate of the flow angle in the impingement region increased.

Figure 9 shows the pressure distribution on the axis of the submerged jet at different times under different Reynolds numbers. As shown in the figure, under the condition of $Re = 11,700$, when $t = 0.01$ s, the pressure of the jet along the axis line increased rapidly from a negative value to a positive value and then decreased to 0; with the increase of t, the pressure was maintained at the value of 0, then decreased, then increased, and finally decreased to 0; compared with Figure 7a, it can be seen that the pressure was the smallest when the velocity was maximum. When the jet velocity distribution remained unchanged, the pressure distribution along the axis of the jet also remained unchanged: The pressure in the free jet region remained near 0, and the pressure in the impingement region increased rapidly to the maximum value and then decreased slightly. With the increase of *Re*, the peak pressure on the axis increased, but its distribution was similar.

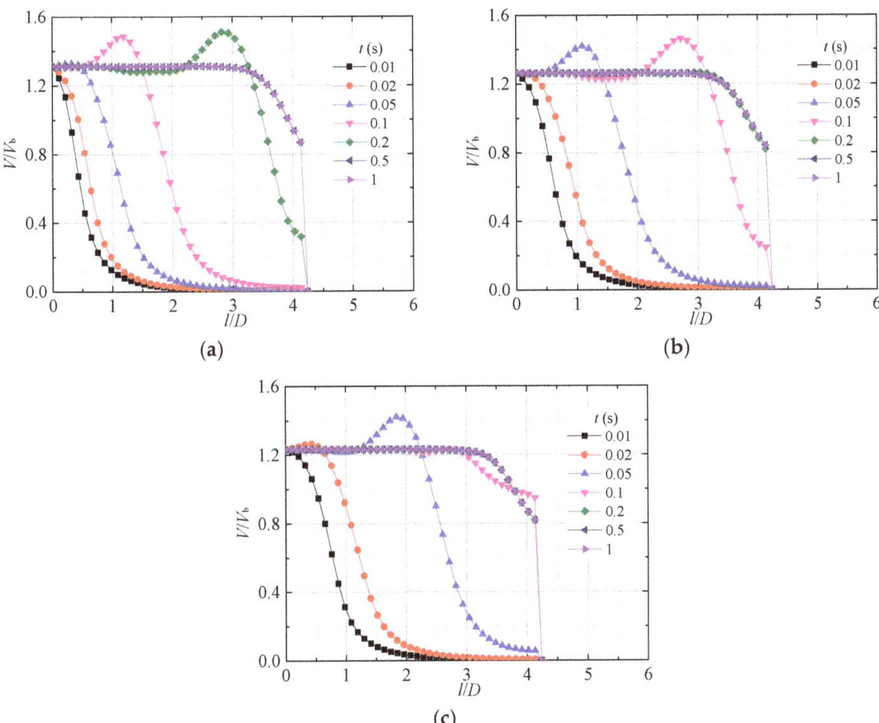

Figure 7. Normalized mean axial velocity V/V_b along the jet centerline. (**a**) $Re = 11{,}700$, (**b**) $Re = 23{,}400$, (**c**) $Re = 35{,}100$.

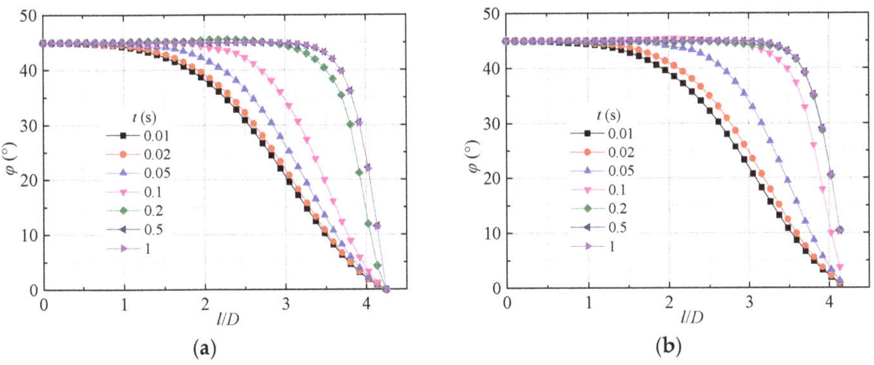

Figure 8. *Cont.*

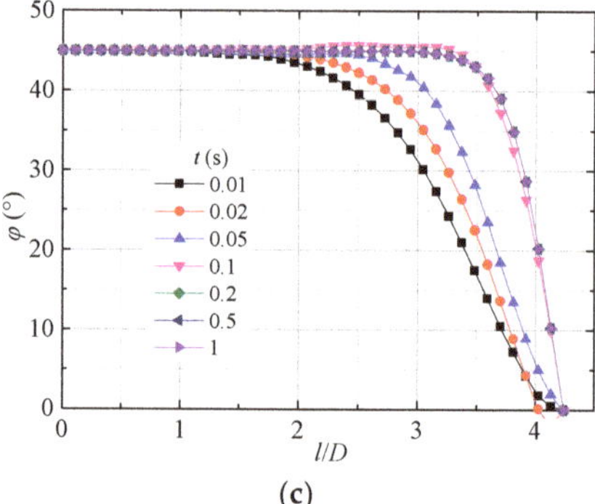

Figure 8. Definition of flow angle φ along the jet centerline. (**a**) $Re = 11{,}700$, (**b**) $Re = 23{,}400$, (**c**) $Re = 35{,}100$.

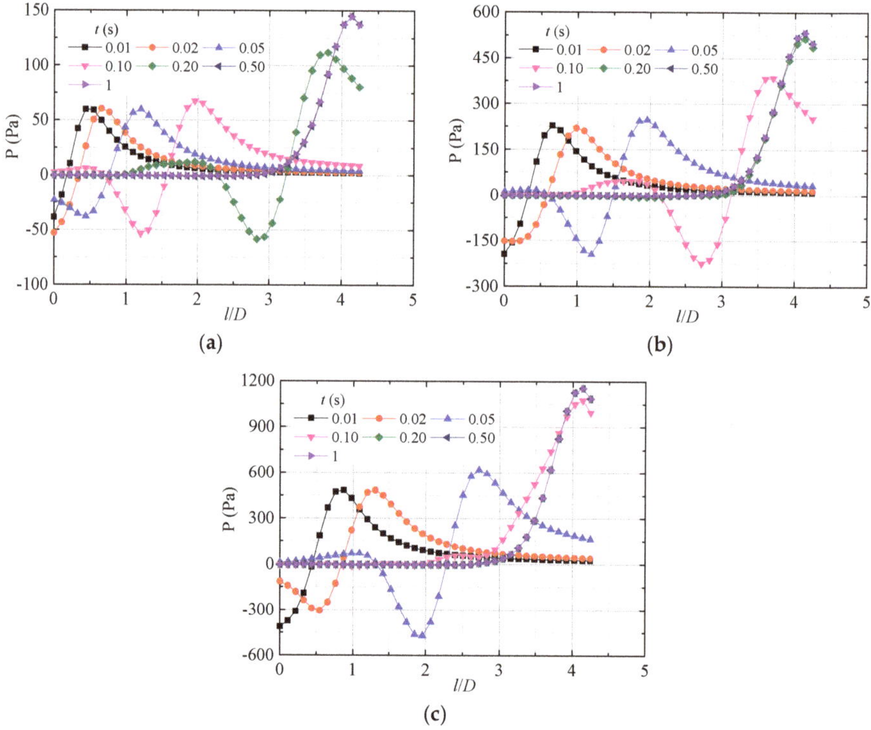

Figure 9. Definition of pressure along the jet centerline. (**a**) $Re = 11{,}700$, (**b**) $Re = 23{,}400$, (**c**) $Re = 35{,}100$.

3.3. Pressure Distribution on the Impingement Plate at Different Times

Figure 10 shows the pressure contours ($Re = 23{,}400$) of the jet impinging on the plate (*oxy* plane) at different times. It can be seen from the figure that when $t = 0.06$ s, the pressure on the impingement plate was almost 0; when $t = 0.07$ s, the pressure on the impingement plate was elliptical and had good symmetry; when $t = 0.08$ s, the pressure increased near the impinging origin. With the increase of time, the maximum pressure on the impingement plate increased gradually. When $t \geq 0.5$ s, the pressure distribution near the impinging origin was similar.

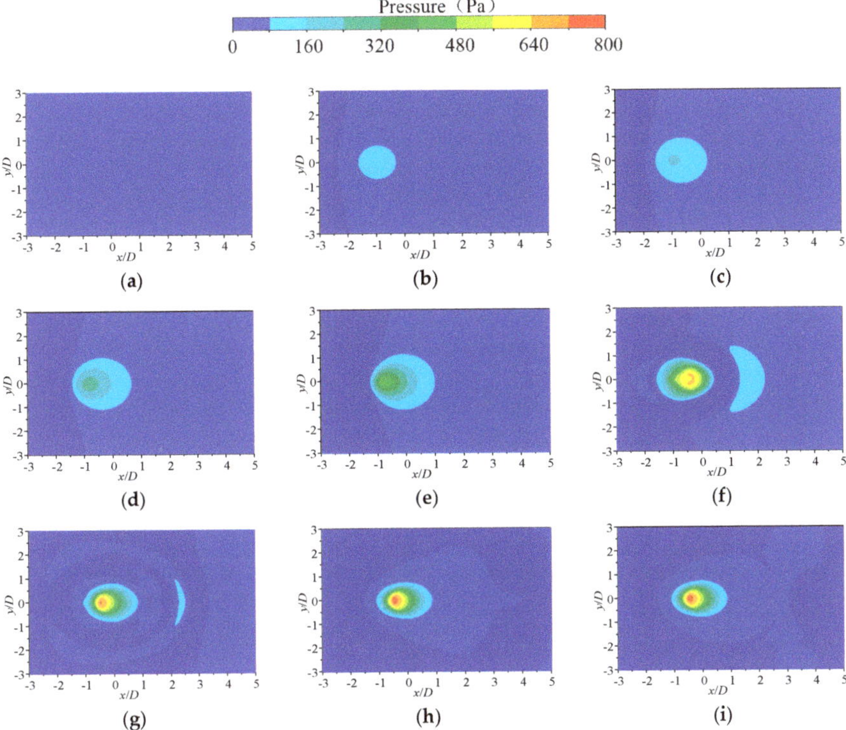

Figure 10. Pressure contours of jet impinging on the plate (*oxy* plane) at different times: (**a**) $t = 0.06$ s, (**b**) $t = 0.07$ s, (**c**) $t = 0.08$ s, (**d**) $t = 0.09$ s, (**e**) $t = 0.10$ s, (**f**) $t = 0.15$ s, (**g**) $t = 0.20$ s, (**h**) $t = 0.20$ s, (**i**) $t = 1.00$ s.

Figure 11 shows the change in the maximum pressure coefficient on the impingement plate with time under different Reynolds numbers. It can be seen from the figure that when $Re = 11,700$, with the increase of time, the maximum pressure coefficient C_{pmax} increased slowly, then rapidly increased, then slowly increased, and finally remained unchanged. With the increase of the Reynolds number, the change rule of the C_{pmax} value with time was similar, but the increasing rate was larger, the maximum value of C_{pmax} decreased slightly after remaining stable, but the time required to reach the maximum value decreased.

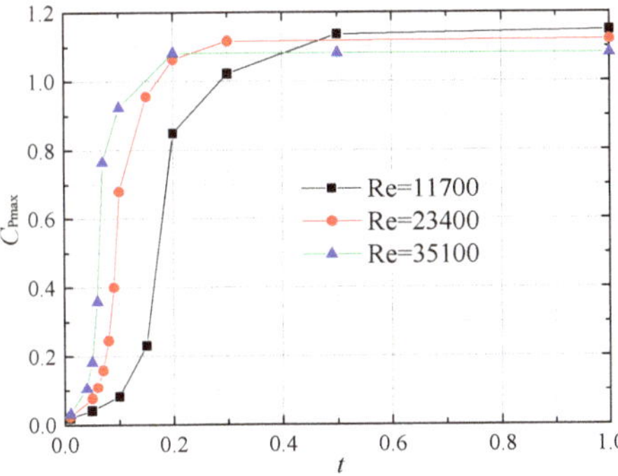

Figure 11. Variation of C_{pmax} with t at different Reynolds numbers.

4. Conclusions

Unsteady numerical calculation of an oblique submerged impinging jet was carried out by using the W-A model, and the following conclusions were obtained:

(1) The jet fluid diffused gradually after leaving the nozzle, the ambient fluid was continuously sucked into the jet, and there were symmetrical vortices around the main body of the jet; when impinging on the plate, most of the jet fluid deflected in the forward flow direction, and the vortex range increased in the forward flow direction, while decreasing in the backward flow direction.

(2) In the process of jet movement, there was a peak value of velocity on the axis, and the pressure at the peak value was the minimum value; with the increase of time, after a certain time, the velocity, flow angle, and pressure distribution along the axis remained unchanged: In the free jet region, the velocity, flow angle, and pressure remained unchanged; in the impingement region, the velocity and flow angle decreased rapidly, while the pressure increased rapidly.

(3) The variation of the maximum impact pressure coefficient on the impingement plate with time was affected by the Reynolds number, but its distribution law was the same.

Author Contributions: Data curation, C.W. and L.C.; formal analysis, W.J. and D.Z.; methodology, L.C. and W.J.; writing—original draft, W.J. and D.Z.; writing—review and editing, T.W. All authors have read and agreed to the published version of the manuscript.

Funding: This research was funded by the National Natural Science Foundation of China (No. 51979240 and 51609105), the Priority Academic Program Development of Jiangsu Higher Education Institutions (Grant number PAPD), the Open Research Subject of Key Laboratory of Fluid and Power Machinery (Xihua University), and the Ministry of Education (szjj2016-061).

Conflicts of Interest: The authors declare no conflict of interest.

References

1. Chaudhari, M.; Puranik, B.; Agrawal, A. Heat transfer characteristics of synthetic jet impingement cooling. *Int. J. Heat Mass Transf.* **2010**, *53*, 1057–1069. [CrossRef]
2. Babic, D.; Murray, D.B.; Torrance, A.A. Mist jet cooling of grinding processes. *Int. J. Mach. Tools Manuf.* **2005**, *45*, 1171–1177. [CrossRef]
3. Jiang, C.; Han, T.; Gao, Z.; Lee, C.H. A review of impinging jets during rocket launching. *Prog. Aeronaut. Sci.* **2019**, *109*, 100547. [CrossRef]
4. Gutmark, E.; Wolfshtein, M.; Wygnanski, I. The plane turbulent impinging jet. *J. Fluid Mech.* **1978**, *88*, 737–756. [CrossRef]
5. Adane, K.F.K.; Tachie, M.F. Experimental and numerical study of laminar round jet flows along a wall. *J. Fluids Eng.* **2010**, *132*, 101203. [CrossRef]
6. Kim, M.; Karbasian, H.R.; Yeom, E. Transient three-dimensional flow structures of oblique jet impingement on a circular cylinder. *J. Vis.* **2018**, *21*, 397–406. [CrossRef]
7. Mishra, A.; Yadav, H.; Djenidi, L.; Agrawal, A. Experimental study of flow characteristics of an oblique impinging jet. *Exp. Fluids* **2020**, *61*, 1–16. [CrossRef]
8. Shi, L.; Zhu, J.; Tang, F.; Wang, C. Multi-Disciplinary Optimization Design of Axial-Flow Pump Impellers Based on the Approximation Model. *Energies* **2020**, *13*, 779. [CrossRef]
9. Wang, H.; Long, B.; Wang, C.; Han, C.; Li, L. Effects of the Impeller Blade with a Slot Structure on the Centrifugal Pump Performance. *Energies* **2020**, *13*, 1628. [CrossRef]
10. Shi, L.; Zhang, W.; Jiao, H.; Tang, F.; Wang, L.; Sun, D.; Shi, W. Numerical simulation and experimental study on the comparison of the hydraulic characteristics of an axial-flow pump and a full tubular pump. *Renew. Energy* **2020**, *153*, 1455–1464. [CrossRef]
11. He, X.; Zhang, Y.; Wang, C.; Zhang, C.; Cheng, L.; Chen, K.; Hu, B. Influence of Critical Wall Roughness on the Performance of Double-Channel Sewage Pump. *Energies* **2020**, *13*, 464. [CrossRef]
12. Wang, H.; Long, B.; Yang, Y.; Xiao, Y.; Wang, C. Modelling the influence of inlet angle change on the performance of submersible well pumps. *Int. J. Simul. Model.* **2020**, *19*, 100–111. [CrossRef]

13. Ghiti, N. Large Eddy Simulation of Three Dimensional Impinging Jets. *Comput. Model Eng. Sci.* **2014**, *99*, 195–208.
14. So, H.; Yoon, H.G.; Chung, M.K. Large eddy simulation of flow characteristics in an unconfined slot impinging jet with various nozzle-to-plate distances. *J. Mech. Sci. Technol.* **2011**, *25*, 721–729. [CrossRef]
15. Gopalakrishnan, R.N.; Disimile, P.J. CFD Analysis of Twin Turbulent Impinging Round Jets at Different Impingement Angles. *Fluids* **2018**, *3*, 79. [CrossRef]
16. Baghel, K.; Sridharan, A.; Murallidharan, J.S. Numerical study of free surface jet impingement on orthogonal surface. *Int. J. Multiph. Flow* **2019**, *113*, 89–106. [CrossRef]
17. Wei, W.; Xu, W.; Deng, J.; Liu, B. Experimental Study of Impact Pressures on Deep Plunge Pool Floors Generated by Submerged Inclined Jets with Controlled Aeration. *J. Hydraul. Eng.* **2020**, *146*, 04020021. [CrossRef]
18. Tian, Z.; Xu, W.L.; Wang, W.; Liu, S.J.; Dong, J.W. Scale effect of impinging pressure caused by submerged jet. *J. Hydrodyn.* **2005**, *17*, 478–482.
19. Shao, J.R.; Li, S.M.; Liu, M.B. Numerical simulation of violent impinging jet flows with improved SPH method. *Int. J. Comp. Meth.* **2016**, *13*, 1641001. [CrossRef]
20. Wang, H.; Qian, Z.; Zhang, D.; Wang, T.; Wang, C. Numerical Study of the Normal Impinging Water Jet at Different Impinging Height, Based on Wray–Agarwal Turbulence Model. *Energies* **2020**, *13*, 1744. [CrossRef]
21. Zhang, X.; Agarwal, R.K. Numerical Simulation of Fountain Formation due to Normal and Inclined Twin-Jet Impingement on Ground. *Fluids* **2020**, *5*, 132. [CrossRef]
22. Zhang, H.; Chen, Z.; Guo, Z.; Zheng, C.; Xue, D. Numerical investigation on the three-dimensional flow characteristics of unsteady subsonic elliptic jet. *Comput. Fluids* **2018**, *160*, 78–92. [CrossRef]
23. Yang, L.; Li, Y.; Ligrani, P.M.; Ren, J.; Jiang, H. Unsteady heat transfer and flow structure of a row of laminar impingement jets, including vortex development. *Int. J. Heat Mass Transf.* **2015**, *88*, 149–164. [CrossRef]
24. Chung, Y.M.; Luo, K.H.; Sandham, N.D. Numerical study of momentum and heat transfer in unsteady impinging jets. *Int. J. Heat Fluid Flow* **2002**, *23*, 592–600. [CrossRef]
25. Zhang, K.; Chen, S. Numerical simulation of self-excited pulsed cavitation nozzle in three-dimensional unsteady flow. *J. Drain. Irrig. Mach. Eng.* **2018**, *36*, 288–293.
26. Zaafouri, M.H.; Habli, S. Numerical analysis of a plane laminar jet in a pulsed coflow. *Int. J. Fluid Mech. Res.* **2019**, *46*, 89–99. [CrossRef]
27. Han, X.; Wray, T.; Agarwal, R.K. Application of a new DES model based on wray-agarwal turbulence model for simulation of wall-bounded flows with separation. In Proceedings of the 47th AIAA Fluid Dynamics Conference, Denver, Colorado, 5–9 June 2017; p. 3966.
28. Chuan, W.; Xikun, W.; Weidong, S.; Weigang, L.; Tan, S.K.; Zhou, L. Experimental investigation on impingement of a submerged circular water jet at varying impinging angles and Reynolds numbers. *Exp. Therm. Fluid Sci.* **2017**, *89*, 189–198.
29. Fairweather, M.; Hargrave, G. Experimental investigation of an axisymmetric, impinging turbulent jet. 1. Velocity field. *Exp. Fluids* **2002**, *33*, 464–471. [CrossRef]

Article

A Theory for Power Extraction from Passive Accelerators and Confined Flows

Robert Freda [1], Bradford Knight [2] and Siddharth Pannir [1,*]

[1]   Design and Engineering, GenH Inc., Charlestown, MA 02129, USA; robert.freda@genh.co
[2]   Department of Naval Architecture and Marine Engineering, University of Michigan, Ann Arbor, MI 48109, USA
*   Correspondence: Siddharth.pannir@genh.co; Tel.: +1-815-603-4887

Received: 24 July 2020; Accepted: 14 September 2020; Published: 16 September 2020

Abstract: No accepted fluid theory exists for power extraction from unpressurized confined flow. The absence of a valid model to determine baseline uniform power extraction in confined flows creates difficulties in characterizing the coefficient of power. Currently, the primary body of research has been limited to Diffuser Augmented Wind Turbines (DAWTs) and passive fluid accelerators. Fluid power is proportional to the cube of velocity; therefore, passive acceleration is a promising path to effective renewable energy. Hypothetical models and experiments for passive accelerators yield low ideal power limits and poor performance, respectively. We show that these results derive from the misapplication of Betz's Law and lack of a general theory for confined flow extraction. Experimental performance is due to the low efficiency of DAWTs and prior hypotheses exhibit high predictive error and continuity violations. A fluid model that accurately predicts available data and new experimental data, showing disk specific maximum C_P for the confined channel at 38% of power available to disk, is presented. This is significantly lower than the 59% Betz freestream limit yielded by hypothetical models when the area ratio equals one. Experiments and their results are presented with non-DAWT accelerators, where new experimental results exceed C_P limits predicted previously and correlate with the proposed predictive model.

Keywords: climate change; renewable energy; wind power; accelerators; turbines; power extraction; Betz; freestream theory

---

1. Introduction

Power extraction from passive acceleration has been actively studied since the 1950's. Passive accelerators are placed in fluid flows, such as wind or hydro currents, to accelerate fluid velocity and can increase the energy density and availability [1] of the resource. Passive accelerators operate in the unpressurized confined flow regime and constitute the primary body of research for unpressurized confined flow. The unpressurized confined flow regime presents a complex fluid mechanics problem for which there is no valid theory that accurately predicts power extraction. Power extraction from confined flows is becoming a subject of importance to the energy landscape and climate mitigation. A valid theory for power extraction from the confined flow is necessary to quantify, develop, and utilize these resources and technologies.

Research into wind and hydro current passive acceleration provides a unique body of experimental work for the development of a theory of power extraction from confined flows. Passive wind and current accelerators are confined systems open to the freestream with no addition of energy to the flow from gravity or a combustion or vacuum chamber. Passive accelerators exhibit the fundamental properties of confined flow power extraction, absent the further complexities introduced in pressurized systems. Accurate theoretical prediction of the experimental performance of wind and current accelerators is the first step in developing a theory of power extraction from confined flows. This paper is concerned with developing a valid theoretical framework for the unpressurized or baseline confined condition.

Renewable Energy (RE) generation from solar or standard wind has low power density per m^2, as defined and detailed by V. Smil [2]. RE's power density has likely restrained its impact on Green House Gas emissions and climate change due to the inability to supply significant power at load centers [3], remote viable resources [4], and low availability over most of the globe [5]. The largest energy market share growth since 2000 has been in coal and natural gas, not RE [6–8]. Acceleration could help change that. Improved passive acceleration and extraction could significantly increase RE's capability to offset fossil fuels using accelerated wind and hydrodynamic power generation within urban environments. A complete theory of extraction from accelerated flows is necessary to determine the potential importance of acceleration to the energy landscape and in combatting climate change.

Passive accelerator flow is the flow is diagrammed in Figure 1. Continuity must be satisfied between the inlet and exit, making accelerators difficult to analyze with free stream theories that require unconfined expansion. Betz's Law establishes the limit for free stream turbines [9]. Hypothetical models presented by van Bussel [10], Jamieson [11], and Werle and Presz (WP) [12] referred to subsequently as Averaging Models (AMs), have attempted to use Betz's Law to establish ideal accelerator power limits. The AM's have sought to validate ideal models with experimental data from Diffuser Augmented Wind Turbines (DAWTs). DAWTs are characterized by short, high-angle diffusers, and are inefficient accelerators [13,14]. AMs predict that accelerators in ideal flow have a Coefficient of Power (C_P) limit that is the same as a normal wind turbine of the accelerator's maximum frontal area.

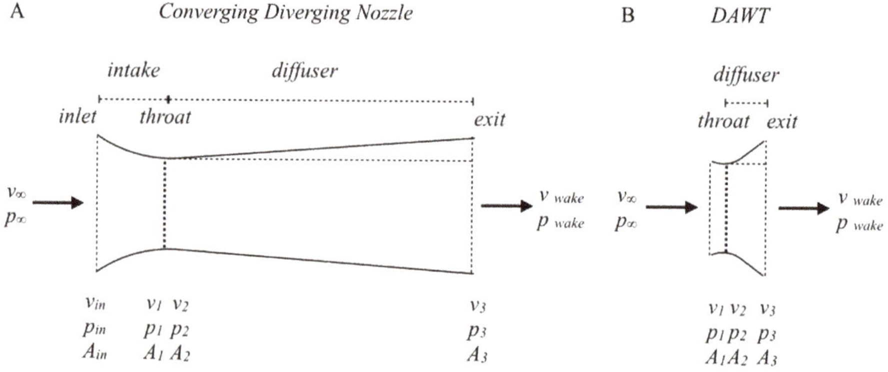

Figure 1. Cross-section schematic of passive accelerator types with stations. (**A**) Converging Nozzle [15]; (**B**) Diffuser Augmented Wind Turbine (DAWT).

A hypothesis, referred to as the Thrust Model, was derived by Knight et al. [15]. This paper provides a detailed mathematical analysis of the preceding AMs and the Thrust model, which was specifically developed for the analysis of power extraction from confined flows, uses accelerator performance and actuator disk body forces to predict C_P performance. We show that AM's misapply Betz's methods by using Betz's velocity averaging without satisfying the necessary assumptions. AMs rely exclusively upon velocity averaging to determine actuator disk conditions, which results in a significant error in the prediction of component variables, C_P results, and violations of continuity.

We show that the Thrust Model is related to Rankine–Froude's Momentum and Actuator Disk Theory [16] and thereby Betz's Law [9] with additional constraints. When related to the Coefficient of Thrust (C_T) and the area ratio between the throat and maximum area is set to one, Rankine–Froude momentum theory's wake velocity equation is the same as the Thrust Model's disk velocity equation. This illustrates the relationship between the Thrust and Rankine–Froude Momentum theory and the additional continuity constraints in the Thrust Model. The Thrust Model and AMs' predictions are compared to experimental data to determine predictive accuracy. New experimental evidence is presented that exceeds the limits predicted by AMs. The Thrust Model accurately predicts all experimental data and indicates significantly higher performance and C_P limits for accelerators than AM limits. All performance and limits herein are related to the power available at the accelerator's maximum area at freestream velocity (P = $0.5\rho A_{max}v_\infty{}^3$).

A Brief History of Accelerators

Interest in accelerators is due to the fact that fluid power increases as the cube of velocity. If one can passively increase the velocity one should be able to extract more power from the flow. This conclusion, based on the standard velocity form of the fluid power equation, has always been at odds with certain assumptions in fluid dynamics. Some suggest that an accelerator cannot extract more power than a free stream turbine of the accelerator's maximum area. Others have suggested this limit could be exceeded under certain conditions which will be examined in the technical analysis in Appendix A. Theorists have formulated hypotheses showing the limit of the maximum area and some have provided potential mechanisms to exceed that limit. Prior research has been unsuccessful in showing that accelerators can increase the power extractable from a fluid related to an equivalent freestream turbine of the accelerator's maximum area. However, this research has been restricted to a single type of accelerator, the Diffuser Augmented Wind Turbine (DAWT). Theories to date have relied strictly on data from this single accelerator type. Theorists such as van Bussel, Jamieson, and WP have used this data to validate ideal theories and limits for accelerators. Problematically DAWT's are not ideal accelerators.

DAWT's have been researched in the field and wind tunnels since the 1970's. Initial research was performed by Igra of Ben Gurion University in Israel [13]. Igra was the first scientist to design a DAWT. The design drivers for Igra's DAWT were to reduce both the accelerator material costs and the swept area occupied by the machine under rotation for different incoming flow directions. With single accelerators under rotation, unless the diffuser length from the point of rotation is less than or equal to the radius of the exit, the effective free stream wind turbine that the accelerator has to outperform is significantly larger than just the maximum frontal area of the accelerator. Research was continued by a team from Grumman Aerospace led by K. Foreman [17]. The Grumman team's work, which culminated in the Vortec 7 [18] in early 2000, constitutes one of the more complete bodies of experimental data. Work was done with both screens and rotors which makes the research particularly useful for deriving limits for accelerators based on Rankine–Froude's actuator disk theory. Figure 2A shows the Grumman device [17] from 1979 that has provided the data for predictive comparisons in this and other papers. Figure 2B,C show the Vortec 7 (B) [18] from 2000 and Flodesign/Ogin (C) [19] from 2015. Historically, high expansion angle diffusers with either a single annulus or a double annulus have characterized DAWTs. Phillips [13] and Wind Lens [20] added a brim at the exit of the diffuser. Using the unsteady effects caused by the brim, the Wind Lens DAWT is the most efficient, with a frontal area C_P of 0.54 [20].

DAWT design has generally failed to achieve the two design goals of its creator Igra [21]. While the shorter diffuser uses less material, the diffuser also experiences significantly higher drag than a freestream turbine which in turn increases structural costs. DAWTs require greater power production to achieve a reduction in the freestream wind turbine Levelized Cost of Electricity. While the DAWT solves the rotational issue, high-expansion angle diffusers perform poorly on an accelerative basis, effectively negating the benefit. Reid's National Advisory Committee for Aeronautics (NACA) work from 1951 [22] clearly shows the reduction in diffuser efficiency that accompanies an increased diffuser expansion angle. Foreman specifically mentions the tradeoff between cost reduction and acceleration [17]. The loss of accelerative efficiency is particularly problematic when DAWTs are used to validate hypothetical limits for accelerators.

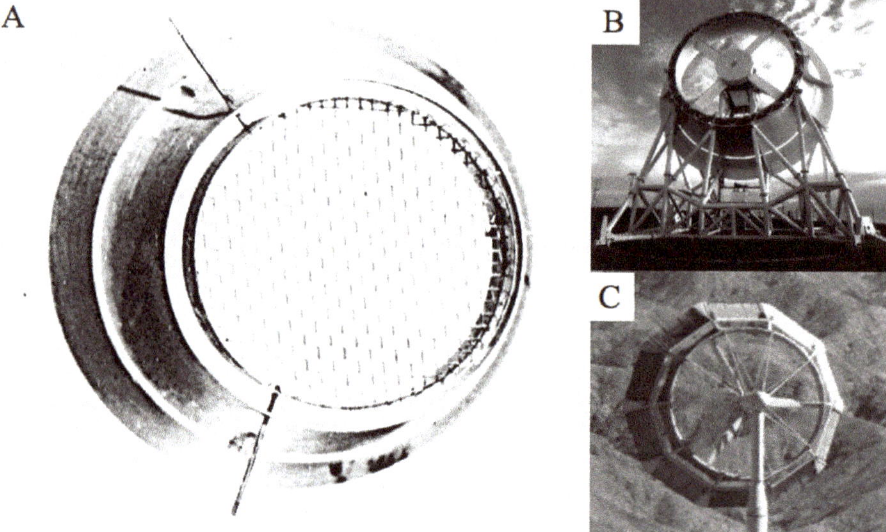

Figure 2. (**A**) Grumman Double Annulus, Boundary Layer Control, 30 Diffuser Augmented Wind Turbines (DAWTs) with wire mesh installed for wind tunnel testing, [17]; (**B**) Vortec 7 [18]; (**C**) Flodesign/Ogin MEWT [19].

The issue of DAWT performance and the conclusions that can be drawn from DAWT research should be understood in the simplest of terms, absent the complexities addressed in the later sections of this paper. A rotor or actuator disk at the DAWTs throat area will experience an accelerated velocity. Using the Grumman data presented in Figure 3 below, the Grumman DAWT [13,17] had an acceleration of $1.56v_\infty$ for a 2.78 area ratio with nothing at the throat. Absent other effects, an increase of wind speed to $1.56v_\infty$ provides power increase to that area of 1.56^3, by the fluid power equation. Independent of further constraints this relation characterizes the most power that the Grumman DAWT actuator disk could possibly produce. Unless the throat area power available to the actuator disk at the accelerated velocity is significantly higher than the power available to an actuator disk of the DAWTs maximum area operating at v_∞, it is highly unlikely that with the additional accelerator constraints the DAWT's actuator disk could exceed its maximum area limit. Figure 3A shows the power available to a disk of the throat area as a function of accelerative efficiency for different area ratios. The values for the Wind Lens and Flodesign/Ogin devices are estimates based on published or claimed C_P. These estimates may contain errors due to the lack of relevant variables in the published data. Error bars on these estimates are provided which assume that the extractor in question operates at a minimum of 0.2 C_{Pdisk} and a maximum of 0.4 C_{Pdisk}. Figure 3B is a detailed image of DAWT performance. Figure 3 demonstrates that all DAWTs to date have operated at less than 70% baseline accelerative efficiency. The lack of efficiency leads to a reduction in fluid power available at the throat. Unsurprisingly, DAWTs have failed to exceed the Betz limit of their maximum frontal area and that failure cannot be considered evidence of a limit or the potential value of acceleration. More efficient accelerators than DAWTs are necessary to realize the benefits of acceleration.

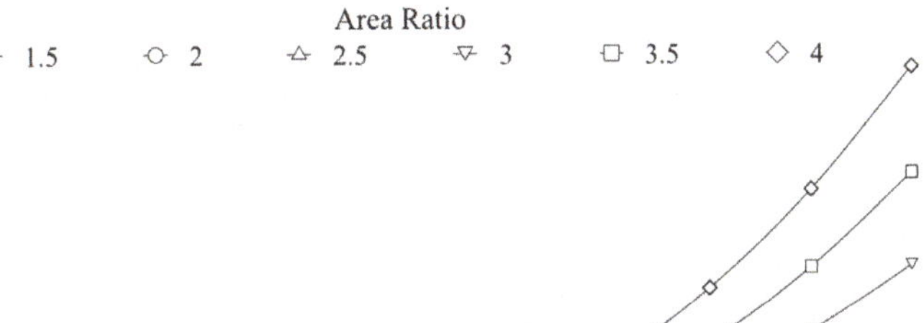

B: Available Power vs. Efficiency for Different Area Ratios from DAWT velocity results

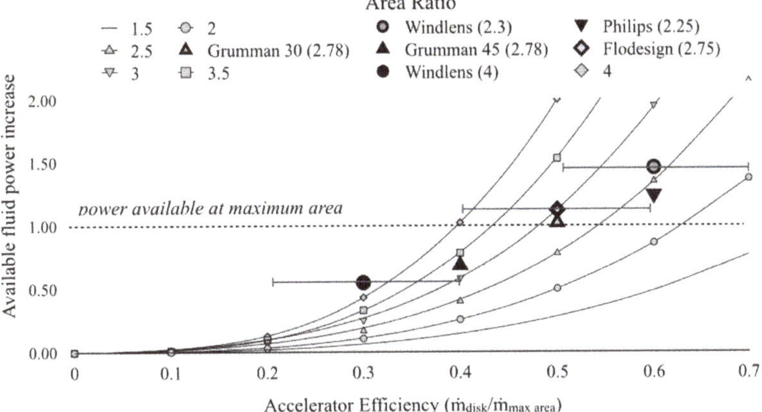

Figure 3. Baseline DAWT Fluid power at throat related to maximum area power ratio. Estimates are based on published/claimed C_P. These estimates are may contain errors due to incomplete reporting of performance. (n) Denotes Area Ratio. Phillips [13], Grumman [17], Wind Lens [20], Flodesign/Ogin [19].

2. Materials and Methods

2.1. AMs: The Inapplicability of Rankine–Froude's and Betz's Averaging Methods to Accelerators

Betz's Law is deemed the power limit for freestream turbines. Betz used averaging of wake and freestream velocities to determine the resistance, velocity, and power extracted by Rankine–Froude's ideal actuator disk. AMs have used this same averaging method to incorrectly predict limits for actuator disks inside of accelerators. Averaging is not valid within accelerators since the flow conditions at the disk cannot be determined by averaging freestream and wake velocities. Betz's Law requires that all flow effects are entrained in the relevant streamtube and that the flow is free to expand at the disk plane. Neither of these requirements are satisfied by AMs' application of Betz's methods.

2.1.1. Overview of Betz's Law for Freestream Turbines

Betz's Law defines the maximum extractable power for a freestream turbine. Betz's equations, [9], utilize averaging of freestream and wake conditions to determine the power of a turbine. Equation (1) shows how Betz averages the wake and freestream velocity to determine the velocity at an actuator disk. The uniform resistance of the disk slows the velocity through the actuator disk. The rate of velocity reduction is represented by axial induction in Equation (2). The resistance for a given C_T is applied at the wake and results in a reduced velocity at the disk due to averaging with v_∞. Therefore, the streamtube must be free to expand around the disk to satisfy mass flow conservation.

$$v_{disk} = \frac{1}{2}\left(v_\infty + v_{wake}\right) \tag{1}$$

$$a = \frac{v_\infty - v_{disk}}{v_\infty} \tag{2}$$

Betz's pressure drop is defined as a function of the axial induction or freestream velocities and uniform density as shown by Equation (3). Betz's Law predicts power as a function of area, velocity, and pressure drop as shown by Equation (4). In Betz's Law, the C_P and C_T can be represented explicitly or in terms of axial induction as shown by Equations (5) and (6).

$$\Delta p = \left(\left(\frac{1}{2}\rho v_\infty^2\right) - \frac{1}{2}\rho\left(\frac{1}{2}\left(v_\infty + v_{wake}\right)\right)^2\right) - \left(\left(\frac{1}{2}\rho v_{wake}^2\right) - \frac{1}{2}\rho\left(\frac{1}{2}\left(v_\infty + v_{wake}\right)\right)^2\right) = 4a(1-a)\frac{1}{2}\rho v_\infty^2 \tag{3}$$

$$P_{\text{Betz}} = v_{disk}\Delta p A_{disk} \tag{4}$$

$$C_P = 4a(1-a)^2 \tag{5}$$

$$C_T = 4a(1-a) \tag{6}$$

As Betz's Law was developed for freestream turbines, it is a fundamental requirement that the extractor cannot be confined. Figure 4 shows the streamlines for a freestream turbine. The streamline shown in black depicts the streamtube which passes through the disk area at the disk velocity as specified by Betz's Law. The light gray lines show the streamtubes for different reference areas. The swept area referenced to v_∞ passes to the outside of the actuator disk area displaying the free expansion requirement inherent to the method. For all streamlines, the area at the disk is greater than the corollary freestream area as v_∞ is greater than v_{disk}. Expansion at and around the actuator disk is a function of Betz's averaging equations as $v_{wake} < v_{disk} < v_\infty$, and $A_{wake} > A_{disk} > A_\infty$.

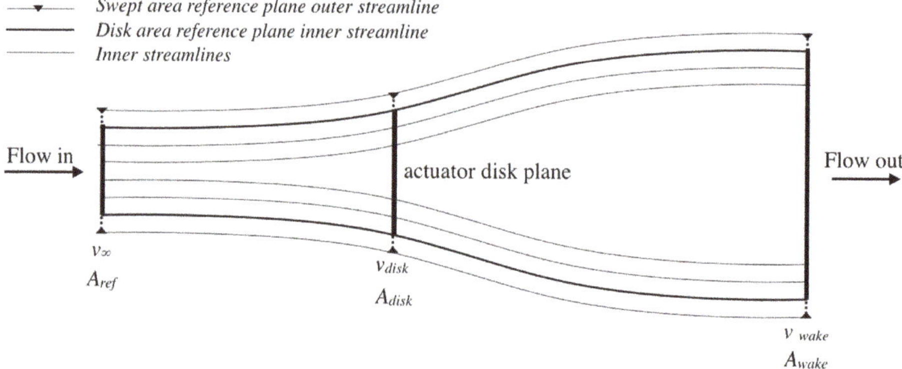

Figure 4. Betz streamlines.

2.1.2. Concerning Rankine–Froude's and Betz's Velocity Averaging Methods and Confined Flows

Betz requires constant expansion of the fluid area in the axial direction to the lowest velocity in the wake. Betz's Law was derived for freestream power extraction, not for confined flow power extraction. In a confined condition, the boundary is closed, and the mass flow area cannot expand at the disk. The mass flow must be conserved throughout the confinement. The resistance at the disk determines the mass flow through the confined channel. Mass flow is rejected at the intake of the confinement instead of at the disk plane as in the freestream case. Confinement engenders a compound reductive effect on the disk velocity which cannot be accounted for by an averaging method.

As can be seen in Figure 4 above, all Betz streamlines are parallel. While not a stated assumption of Betz' control volume, velocity averaging methods will only yield parallel streamlines except at the plane of symmetry as all v_∞ reference areas produce the same results related to C_P. A confined disk cannot have universally parallel streamlines, expanding from all reference areas to the rotor to the wake. Figure 5 depicts a simple tube confinement that is open on either end and has an actuator disk in the middle. For a confined disk resistance is experienced at the actuator disk and the flow expansion occurs at the intake. Therefore, the reference area streamlines diverge at the intake and cannot be parallel. The streamlines inside the tube are not free to expand. It is stated as a requirement of Betz' Law that all relevant flow effects be entrained at the actuator disk [9], which is not satisfied for the confined case by the averaging method. In confined flow, averaging cannot satisfy the assumptions that justify its application. Averaging is therefore incapable of determining disk velocity in confined conditions.

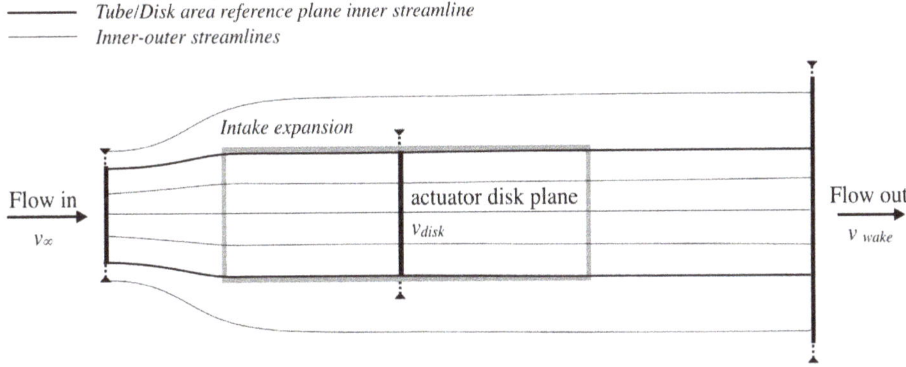

Figure 5. Streamlines around actuator disk inside of a tube.

2.1.3. AMs' Misapplication of Betz's Law and Velocity Averaging

The AMs, proposed by van Bussel [10], Jamieson [11], and WP [12], all rely on the Betz' averaging methods to predict actuator disk velocity and pressure drop in the accelerator channel. The AMs assume that the resistance of a disk in a confined channel is the same as the resistance of the disk in the freestream. The AMs present no derivation, argument, or experimental evidence to justify or validate this assumption. While AMs all use different equation development, the averaging method is fundamental to all AMs irrespective of the apparent differences in variables. The specific relation of each AM to the general averaging method is addressed in Appendix A.

Using velocity averaging for confined flows creates fundamental control volume and conservation problems. Experimental evidence presented in Appendix A will further show that the rate of resistance for a free stream rotor and the streamline expansion at the disk is not the same as the rate of resistance for a confined rotor and the streamline expansion at the intake. The general AM control volume is stated to be the volume depicted in Figure 6 by van Bussel [10], Jamieson [11], and WP [12].

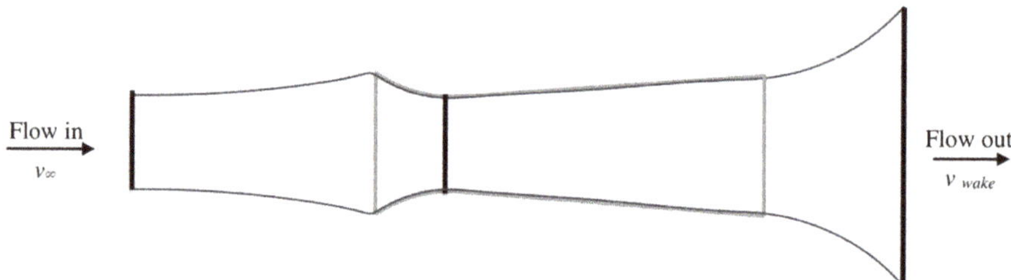

Figure 6. Control Volume described by Averaging Models (AMs).

The stated control volume does not expand at the disk and therefore averaging cannot entrain the relevant flow effects at the actuator disk as noted in the discussion above. Furthermore, under extraction, axial induction is present. Mass is rejected at the intake in the AM control volume, but the mass flow rate external to the nozzle in the AM treatment is associated with external averaging not the applied resistance at the throat. Van Bussel's, Jamieson's, and WP's control volumes violate Betz's assumptions. All flow effects are not entrained in the averaging equations and the flow is not able to expand at the actuator disk.

Additionally, the presented AM control volume is not the only control volume applied in some AMs. Some of the AMs actually use two control volumes to satisfy the assumed energy limits. AM theorists begin with the assumption that the only energy available to do work is the kinetic energy available at v_∞ for the accelerator's maximum area. Van Bussel specifically states, in his 1999 paper [23] that "the amount of energy that can be extracted from the flow per unit of volume is the same as for an ordinary wind turbine". This assumption is reiterated with a second assumption which van Bussel states in his 2007 paper [10], "it is assumed that at the exit of the diffuser the same conditions apply as just after an ordinary wind turbine (assuming no extra back pressure)". To satisfy these assumptions, the AMs must further adapt Rankine–Froude's and Betz's equations resulting in the application of two control volumes and continuity violations.

Equations (7)–(10) are the explicit and axial induction forms of the general AM equations. Power is defined in Equation (7). Equation (10) has Equations (8) and (9) substituted for v_{disk} and Δp, respectively.

$$P_{AM} = v_{disk}\Delta p A_{disk} \tag{7}$$

$$v_{disk} = \frac{1}{2}(\beta v_\infty + \beta v_{wake}) = \frac{1}{2}\beta(v_\infty + v_{wake}) = \beta(1-a)v_\infty \tag{8}$$

$$\Delta p = \frac{1}{2}\rho\left(v_\infty^2 - v_{wake}^2\right) = \frac{1}{2}\rho(4a(1-a))v_\infty^2 \tag{9}$$

$$P_{AM} = \frac{1}{4}\beta\rho(v_\infty + v_{wake})\left(v_\infty{}^2 - v_{wake}{}^2\right)A_{disk} = \frac{1}{2}\rho\beta v_\infty^3 4a(1-a)^2 \tag{10}$$

Equation (8) shows that β is distributive in the velocity case and can be applied directly to the averaged velocity value or its components. Averaging is inherent in the axial induction form of the equation. There is no reduction from the pure application of β to the averaged velocity which shows the absence of the mass flow constraints necessary to model confined flow. There is also no application of β in Equation (9), the pressure drop equation. Given the initial assumption that the accelerator maximum area is the kinetic energy limit of the accelerator and the absence of mass flow constraints in the v_{disk} equation, the removal of β from the pressure drop equations is necessary to satisfy the AMs' kinetic energy assumption. The AM power as defined by Equations (7) and (10) contains the error found in Equation (8).

Jamieson and WP start with the axial induction form of the equations and note that β drops out of the equations. Van Bussel specifically justifies the removal of β as v_{disk} drops out of Betz's pressure drop equations. β cannot be considered distributive in Equation (9), due to the square of velocities. Therefore, the application of β within the squared term or outside of the squared term must be justified and validated experimentally. AMs' exhibit high predictive error for v_{disk} and Δp when compared to experimental results, as will be shown in Section 4. This indicates that there is no basis for van Bussel's removal of the β variable or any application β outside the squared term. In the v_{disk} calculation, β is applied to both v_{∞} and v_{wake}. If applied consistently with one application of β in the v_{disk} equation the result is Equation (11).

$$\Delta p = \frac{1}{2}\rho((\beta v_{\infty})^2 - (\beta v_{wake})^2) \tag{11}$$

Van Bussel states that this cannot be the case, "The maximum achievable power however is not equal to $\beta^3\gamma^3 C_{Pmax}$, where C_{Pmax} is the maximum achievable power of the wind turbine without diffuser" where γ is backpressure. Van Bussel is in part correct, but only because the consistent application of β has no mass flow constraints on the free expansion of the fluid around the disk and will over-predict accelerator power by a large margin. The β^2 result is the same as applying β to the standard fluid power equation within the cubed term and then correcting for the maximum area used.

In the consistent application of β, the error is limited to an incorrect control volume which ignores the confinement. The over-prediction of both v_{disk} and pressure drop could be considered a limit, albeit one that can never be achieved.

The application used by the AM theorists is more problematic. It results in a second control volume, beyond the stated control volume, which simultaneously leads to over-prediction of v_{disk} and under-prediction of pressure drop. The AM's misapplication of β leads to two different velocities being applied to the actuator disk at the same time neither of which reflects the relevant flow effects and conditions.

The AMs' utilized control volumes for velocity and pressure drop are shown in Figure 7 below. The AMs assert that for an ideal accelerator the pressure drop is that of the exit plane area and freestream velocity. The AMs, therefore, disregard the fact that the pressure drop occurs in the confinement of the throat of the accelerator and at the throat velocity. The top image of Figure 7 shows the axial induction form of the pressure drop in the AM equations for an ideal accelerator and the absence of an accelerator in the equation's control volume. The lower image in Figure 7 depicts the control volume which the AMs claim to use. The velocity at the throat of the lower control volume is combined with the pressure drop from the upper control volume to determine the AM power. The control volume in the lower image also has ideal assumptions applied to it. It is unlikely that for an accelerator with an inlet of the same size as the exit that the flow would stay attached to the outside of the accelerator as shown by the light grey line or that the confined rotor would have the same velocity as a freestream rotor increased by β.

2.1.4. The Tube Case (an Accelerator Where $\beta = 1$)

The problems inherent in the general AM are most evident in the application of AM methods to the tube case where $\beta = 1$. Certain AM authors [11,12] claim that their methods provide a general theory for extraction from either freestream or confined flows. A general theory for power extraction for a confined flow should be valid for all positive values of β, 1, or greater. For an infinitely thin-walled tube, the diameter of the tube and the actuator disk are the same. When $\beta = 1$ the AMs yield the Betz limit of the tube area. This is the same as a freestream turbine in the same area. When an extractor is placed in a tube, the extractor's performance is lower than it would be in the free stream.

Figure 8A depicts a RANS CFD study of a two-bladed Horizontal Axis Wind Turbine (HAWT), without a hub, operating in the freestream and a long tube. Figure 8B shows that the power reduces from the initial freestream power in a tube. Images of the flow are RANS CFD velocity plots, which bisect the rotor. Red is high velocity and blue is low velocity. High velocity is shown in the center of the turbine where a hub is not present and in the tip vortex due to a 15% tip gap.

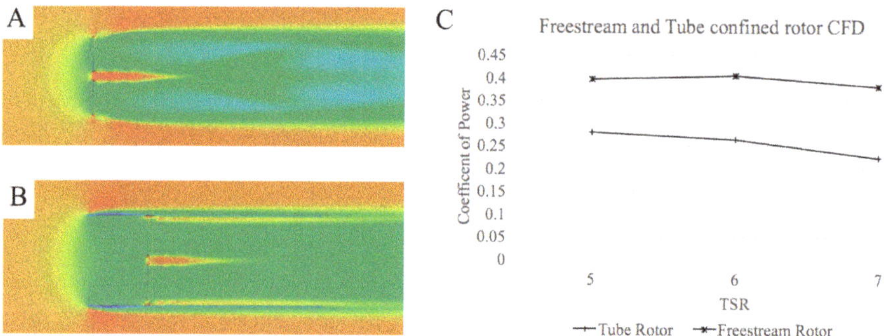

Figure 7. AM Pressure Drop and Rotor Velocity Equation streamlines for ideal accelerators.

Figure 8. Turbines in (**A**) free stream and (**B**) tube; (**C**) power vs. TSR for each case.

As intuition and Figure 8 shows, a wind turbine (or actuator disk) inside of a tube operates at a lower performance than a freestream turbine (or freestream actuator disk) due to the reductive effect of the confinement. The AMs assert that a turbine will have the same performance in a tube as in the freestream. Contrary to the AMs, analytical models, like that presented by Lawn, predict that the power in a tube will be less than in the freestream [24].

Figure 8 shows that mass is rejected at the intake of the tube and the wake begins to expand at the intake. The AMs predict that the tube is collapsible into a single plane through invalid assumptions, which removes the confinement of the tube.

Figure 9 depicts the AM streamtubes when $\beta = 1$, based on the equations for disk velocity and pressure drop, overlaid with the RANS CFD long tube case with the two-bladed HAWT. The dimensions of the tube are dashed. The top image of Figure 9 depicts the stated stream tubes of the AMs for a tube. However, the result of the AMs' equations is that the entire volume of the tube can instead be collapsed into a single extraction plane. For an ideal tube, the AMs predict that the power extraction inside of a tube will be the same as for a freestream turbine. The bottom image of Figure 9 depicts the streamlines of the result of the AMs. These AM streamlines based on the axial induction equations for this condition are the same as the Betz streamlines as there is no reductive effect caused by the confinement. Figure 9 demonstrates that the AM equations do not include a reductive effect caused by the confinement and that the AMs fundamentally ignore the presence of an accelerator. The AM streamtubes for the $\beta = 1$ case is the same for both velocity and pressure drop, unlike the cases shown above in Figure 7 for $\beta > 1$.

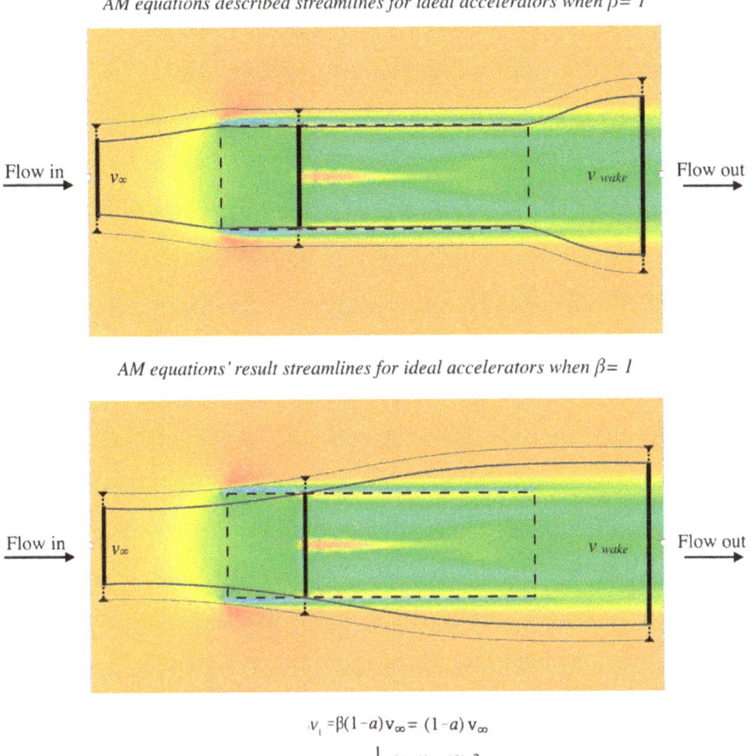

$$v_1 = \beta(1-a)v_\infty = (1-a)v_\infty$$

$$p_1 - p_2 = \frac{1}{2}\rho[4a(1-a)]v_\infty^2$$

Figure 9. AM equation streamlines for ideal accelerators when $\beta = 1$.

As will be shown in later sections, the rate of expansion at the intake depends on the resistance at the disk within the confinement, not on the averaged external velocities. While all AMs suffer from the flaws caused by averaging, the particular execution varies slightly between methods.

3. The Thrust Model Derivation and Comparison to Rankine–Froude Momentum Theory

3.1. The Thrust Model

A complete theory for power extraction in accelerators must predict performance for both ideal and non-ideal functions directly from the actuator disk in the confined channel and be valid for all values of β equal to or greater than one. Qualification of a specific accelerator's performance is necessary to model non-ideal states and to benchmark a particular geometry. The Thrust Model [15] satisfies these requirements.

3.1.1. Thrust Model Derivation

Bernoulli's Law, as specified in Equation (12) governs pressure conditions within the accelerator. The Thrust Model assumes that density is constant through the incompressible assumptions. Stations 1 and 2, as shown in Figure 1 are immediately forward and rearward of the throat of the accelerator and by mass conservation, $v_1 = v_2$, and is related to v_∞ by the area ratio, β. β relates A_1 to the greater of A_{in} or A_3 per Equation (13). Equation (14) defines the initial ideal throat velocity, v_{1o} in relation to v_∞, without extraction (C_T equal to 0). Non-ideal accelerator efficiency, r, is defined in Equation (15) and characterizes the baseline efficiency of the accelerator, which qualifies the accelerator's available power without an extractor present. The Thrust Model assumes incompressible flow.

$$p_{tot} = p_\infty + \frac{1}{2}\rho v_\infty^2 = p_{in} + \frac{1}{2}\rho v_{in}^2 = p_1 + \frac{1}{2}\rho v_1^2 = p_2 + \frac{1}{2}\rho v_2^2 = p_3 + \frac{1}{2}\rho v_3^2 = p_{wake} + \frac{1}{2}\rho v_{wake}^2 \tag{12}$$

$$\beta = Max\left(\frac{A_{in}}{A_1}, \frac{A_3}{A_1}\right) \tag{13}$$

$$v_{1o} = \beta\, v_\infty \tag{14}$$

$$r = \frac{v_{1o}}{\beta v_\infty} \xrightarrow{yields} v_{1o} = r\beta\, v_\infty \tag{15}$$

The extractor is represented by a uniform porous resistance at the disk per actuator disk theory. The disk's closed porosity is equal to the local coefficient of thrust, C_T, defining the body forces acting on the disk. By mass conservation, the resistance of the extractor must immediately communicate through the accelerator as related to β, as shown by Equation (16). The available thrust in the initial flow and thrust on the disk for a given value of C_T is specified by Equations (17) and (18) respectively. As C_T increases with extraction, the magnitude of v_1 decreases from the initial velocity, and the pressure drop, Δp, across the disk increases.

$$\beta_{in} v_{in} = v_{1o} = r\beta v_\infty = v_2 = \beta_3 v_3 \tag{16}$$

$$F_{1o} = \frac{1}{2}\rho(v_{1o})^2 A_1 = \frac{1}{2}\rho(r\beta v_\infty)^2 A_1 \tag{17}$$

$$F_{1extracted} = F_{1o}C_T = \Delta p_1 A_1 \tag{18}$$

Equation (19) shows the remaining force in the fluid. We define Equation (20) as the result of the substitution of Equations (17) and (18) into Equation (19). Equations (20) and (21) define the remaining force in the fluid referred to v_{1o} and v_1 respectively. $F_{1remaining}$ is set as equal and solved for v_1, resulting in Equation (22) for the velocity in the accelerator under extraction. Force remaining and velocity anywhere in the accelerator are related to v_1 and $F_{1remaining}$ by the local value of β. The resistance of the disk can be characterized at any station related to the local value of β and is related to resistance at the intake by β_{in}. Equation (22) gives the disk velocity related to C_T as C_T varies from zero to one. Equation (23) gives the pressure drop across the extractor. Equation (24) shows the power extracted. Equation (25) shows the system C_P is where A_{max} is the larger of A_{in} or A_3. Equation (26) specifies the disk specific C_P.

$$F_{1remaining} = F_{1o} - F_{1extracted} \tag{19}$$

$$F_{1remaining} = \frac{1}{2}\rho v_{1o}^2 A_1(1 - C_T) \tag{20}$$

$$F_{1remaining} = \frac{1}{2}\rho v_1^2 A_1 \tag{21}$$

$$v_1 = \left((r\beta v_\infty)^2(1 - C_T)\right)^{.5} = (r\beta v_\infty)\sqrt{(1 - C_T)} = \beta_{in} v_{in} \tag{22}$$

$$\Delta p_1 = \frac{1}{2}\rho\left((r\beta v_\infty)^2 - v_1^2\right) = \beta_{in}{}^2 \Delta p_{in} \tag{23}$$

$$P = v_1 F_{1extracted} = \Delta p_1 A_1 v_1 \tag{24}$$

$$C_{P\_system} = \frac{F_{1extracted} v_1}{\left(\frac{1}{2}\rho A_{max} v_\infty^3\right)} \tag{25}$$

$$C_{P\_disk} = \frac{F_{1extracted} v_1}{\left(\frac{1}{2}\rho A_1 v_1^3\right)} \tag{26}$$

The obstruction presented to the flow by the extractor-accelerator pair has a variable area related to a given C_P and C_T of the pair. Rejection of mass flow under operation increases the effective area of the accelerator, akin to increasing the coefficient of drag of a system to a greater value than one. From the reference plane of the flow, the body appears larger than its area. The AMs restrain the energy streamtube to the swept area of the device projected forward; whereas the Thrust Model does not constrain the energy effects. An object like a flat plate has a coefficient of drag greater than one, therefore the object experiences more force than the force present in its swept area projected forward. A similar effect occurs with an accelerator, where the available power for extraction increases from the initial state and swept area of the accelerator to a maximum at $C_T = 0.65$. Equation (27) specifies the effective area of the device calculated from power.

$$A_{effective} = \frac{V_1 F_{extraction}}{\frac{1}{2}(v_\infty + v_{wake})\left(\rho V_\infty^2 - \rho\left(\frac{1}{2}(v_\infty + v_{wake})\right)^2\right)} \tag{27}$$

3.1.2. The Thrust Model's Relationship to Rankine–Froude's Actuator Disk (Momentum) Theory and Betz' Law

Rankine–Froude's actuator disk theory and Betz's Law rely upon velocity averaging to determine disk velocity. Due to velocity averaging the Thrust Model may be unrelated to either and breaks the assumptions on which both are based in terms of power extractable from a flow based on kinetic energy limits. This is not the case. The Thrust Model is an extension of Rankine–Froude's Theory and Betz' Law to confined flows using body forces at the disk instead of averaging external velocities. The Thrust Model observes all assumptions and parameters on which Rankine–Froude's and Betz' models are based. As noted in the AM section, the specific application of β is crucial to the accurate formulation of theory and prediction of experimental behavior.

In a freestream flow, Rankine–Froude's actuator disk relies upon three velocities to derive disk power, incident (v_∞), disk (v_1), and resultant (v_{wake}). Pressure drop is calculated from incident to resultant velocities. The disk velocity is the average of the incident and the resultant velocities. Therefore, power is a function of all three velocities. The pressure drop resistance is applied at v_{wake}, as shown by Rankine–Froude's Equation (4) or equation (28).

$$\Delta p = \frac{1}{2}\rho\left(v_\infty^2 - v_{wake}^2\right) \tag{28}$$

The averaging to produce v_1 introduces the limitation of half the resistance at the disk. A valid general theory for a confined flow should arrest the flow in the channel when $C_T = 1$. For a confined channel the fluid effects of the extraction are fundamentally different than in the freestream case. The "wake" expansion occurs at the inlet and the wake mixing occurs rearward of the exit. In the Rankine–Froude or Betz model, the expansion occurs around the rotor and the mixing occurs rearward of the wake. AMs interpret this condition for confined channels as a pressure drop from v_∞ to v_{wake} independent of the value of β. In confined flow, there is no distinct consequent v_{wake} related to a "wake" area. The v_{wake} condition is coincident with v_1 due to mass flow constraints. The normal maximum wake expansion for a given value of C_T, v_{wake}, is the velocity within the confined channel, which predicates that the relevant pressure drop velocity is βv_{wake}. Rearward of the confinement this reduced exit velocity mixes with the external "wake" from the intake and the freestream as in the Betz model after the point of maximum expansion. Therefore, in confined flow, there are only two relevant velocities, the incident, and resultant. The introduction of averaging to determine disk velocity therefore introduces conservation violations if applied to the confined channel.

A derivation of Rankine–Froude and Betz wake velocities related to C_T shows that the resultant velocity, v_{wake}, the equation is the disk velocity equation of the Thrust Model. When β and r are equal to one, corollary to a tube in the freestream, Equation (10) of the Thrust Model becomes Equation (29).

$$v_1 = v_\infty \left(1 - C_T\right)^{.5} \tag{29}$$

To compare the Thrust Model to the Rankine–Froude and Betz models, the wake velocity must be derived related to C_T. Betz's Law [9] states that the pressure drop across a free stream turbine can be defined by the dynamic pressure variation between the wake and the freestream. This can be related to C_T by pressure drop over freestream dynamic pressure, q_∞. Therefore, v_{wake} referred to C_T can be derived as follows. Equations (30) and (31) describe the pressure conditions of the incident velocity and the resultant velocity.

$$p_\infty + q_\infty = p_\infty + \frac{1}{2}\rho\, v_\infty^2 \tag{30}$$

$$p_\infty + q_{wake} = p_\infty + \frac{1}{2}\rho\, v_{wake}^2 \tag{31}$$

Static terms drop out and C_T is given by Equation (32) and pressure drop, in terms of freestream and wake dynamic pressure, is given by Equation (33).

$$C_T = \frac{\Delta p}{q_\infty} \tag{32}$$

$$\Delta p = q_\infty - q_{wake} \tag{33}$$

Combining Equations (32) and (33) and simplifying results in Equations (34)–(36).

$$C_T = \frac{q_\infty - q_{wake}}{q_\infty} \tag{34}$$

$$C_T\, q_\infty = q_\infty - q_{wake} \tag{35}$$

$$q_{wake} = q_\infty - C_T\, q_\infty \tag{36}$$

Substituting Equations (30) and (31) for pressure and simplifying in Equations (37)–(40) yields Equation (41).

$$\frac{1}{2}\rho\, v_{wake}^2 = \frac{1}{2}\rho\, v_\infty^2 - C_T \frac{1}{2}\rho\, v_\infty^2 \tag{37}$$

$$\frac{1}{2}\rho\, v_{wake}^2 = \frac{1}{2}\rho\, v_\infty^2 \left(1 - C_T\right) \tag{38}$$

$$v_{wake}^2 = \frac{\frac{1}{2}\rho v_\infty^2 (1 - C_T)}{\frac{1}{2}\rho} = \left((v_\infty)^2 (1 - C_T)\right) \tag{39}$$

$$v_{wake} = \left((v_\infty)^2 (1 - C_T)\right)^{.5} \tag{40}$$

$$v_{wake} = v_\infty \left(1 - C_T\right)^{.5} \tag{41}$$

The Rankine–Froude and Betz wake velocity Equation (41) is the same as the Thrust Model disk velocity, shown by Equation (29). This shows that when related to C_T, the Thrust Model throat velocity is the same as Rankine–Froude and Betz wake velocity. At this velocity, the streamtube is no longer expanding and is in a confined state. In the Thrust Model, the disk and wake velocities are coincident. This shows the compound reductive effect of the confinement on disk velocity and power. This equation can be brought to 0 m/s at $C_T = 1$. Figure 10 shows the stream tubes of Rankine–Froude and the Thrust Model and the relevant pressure conditions.

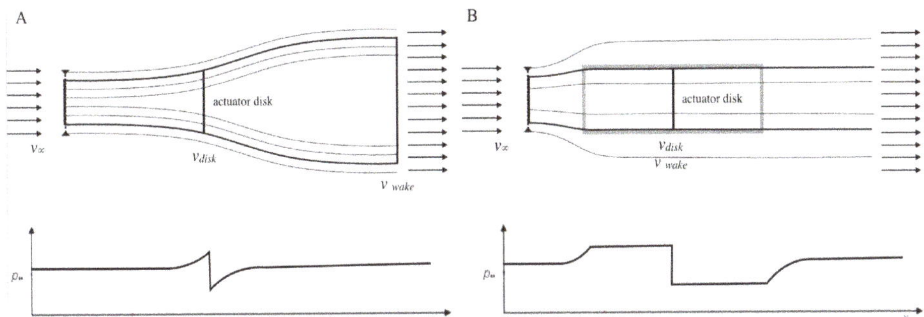

Figure 10. Relevant Pressure conditions: (**A**) Rankine–Froude Actuator Disk Theory and (**B**) Thrust Model.

As Rankine–Froude and Betz are models for freestream turbines, C_T in their models is non-dimensionalized with the freestream velocity. As shown above, the Thrust Model's uses initial throat velocity to non-dimensionalize the pressure drop to calculate C_T, since initial throat velocity is the source of the pressure drop in the Thrust Model. The AMs assert that the pressure drop for a turbine inside of an accelerator occurs in the same way as a freestream turbine. Therefore, AMs use freestream velocity to calculate C_T.

By non-dimensionalizing pressure drop with the relevant velocity for the C_T calculation the Thrust Model, Rankine–Froude's theory, Betz's Law, and the AMs can be compared on the same basis. When non-dimensionalized in this way, all models have an open disk when $C_T = 0$ so that $v_{wake} = v_\infty$. When $C_T = 1$ a singularity occurs for all models. When $C_T = 1$, the Thrust Model has 100% disk solidity so v_1 and mass flow through the accelerator goes to zero, therefore there is a singularity at the intake as there must be infinite mass flow area expansion at the intake. When $C_T = 1$ for Rankine–Froude and Betz there is a singularity at the wake where there is infinite mass flow area expansion as v_{wake} goes to zero. At this point, the axial induction at the disk is $a = 0.5$. The AMs contain the same singularity because they assert that the pressure drop is the same as for a freestream turbine.

The fundamental error of preceding attempts to apply Rankine–Froude's momentum methods to pressure drop in confined flow is the application of β. The Thrust Model pressure drop equation (Equation (22)) can be restated in Rankine–Froude's form as shown by Equation (42).

$$\Delta p = \frac{1}{2}\rho((r\beta v_\infty)^2 - (r\beta v_2)^2) \text{ where incident velocity } = r\beta v_\infty \text{ and resultant velocity } = r\beta v_2 \quad (42)$$

v_2 is the resultant velocity within the channel after the work is done as in the Rankine–Froude and Betz wake. For $\beta = 1$ the disk velocity is lower than the freestream case and the maximum rotor specific C_P is 38% as predicted by the Thrust Model and Lawn's analytical model. This pressure drop correlates to the difference between the accelerated incident and resultant velocities. As will be shown in comparison to experimental data this yields an accurate pressure drop for a confined disk. As disk velocity cannot be the freestream disk velocity, the relevant velocity at the disk must be the resultant velocity in the channel under mass flow constraints.

4. Discussion

4.1. Comparison of Hypotheses and Experimental Data

The primary determinant of limits for accelerator-extractor pairs is the rate of mass flow rejection due to actuator disk resistance at the throat. AM methods have no mechanism to correctly qualify an experimental accelerator. Equations (43) and (44) show that the Grumman accelerator, as referenced by van Bussel, had available fluid power that at $C_T = 0$ was only marginally greater than the fluid power available to a freestream turbine of the accelerator's maximum area [10,17].

$$r_{\text{Grumman}} = \frac{1.56 v_\infty}{2.78 v_\infty} = 0.56 \tag{43}$$

$$\frac{P_{\text{available,0@throat}}}{P_{\text{available, 0 @A}_{\max}}} = \frac{(1.56 v_\infty)^3}{2.78 (v_\infty)^3} \tag{44}$$

As noted in Section 1, any validation based on unqualified DAWT experiments would erroneously confirm a limit of roughly the maximum area. Without proper qualification, non-ideal tests cannot validate limits.

Careful consideration is necessary to compare a system like the Thrust Model with systems like the AMs. We present two methods of comparison. The first method bounds the comparison on the same C_T mapping basis. The second method non-dimensionalizes the thrust on the extractor with the local disk velocity.

AM and Thrust Results Mapped to CT Boundary of 0–1

The C_T mapping basis is derived from the relations in Section 3. As noted in Section 3, the Thrust Model's C_T in non-dimensionalized in terms of incident throat velocity. Furthermore, the Thrust Model's disk and wake velocities are coincident. This provides the numerical basis for non-dimensionalized v_1 to be varied from zero to one. This is the same basis as the Betz, Rankine–Froude, and AM methods. The Thrust Model requires three-dimensional mapping since it has variation in three independent variables, r, β, and C_T. The Betz and Rankine–Froude systems have a single independent variable C_T as defined by Equation (41). As the AMs are an extension of the Betz and Rankine–Froude systems, they are univariate in their C_P and pressure drop equation. However, the AMs are multivariate in the disk velocity equation due to variation in the β dimension as can be seen in Figure 11A. To compare univariate (Rankine–Froude/Betz), multivariate (Thrust Model), and blended univariate/multivariate systems (AMs) it is a useful visualization to set the same boundary conditions for the two and three-dimensional mapping. The boundary for comparison is established by the bounds of the AM disk velocity, shown in Figure 11A, which has the same bounds as the univariate equations. Figure 11B shows the same boundaries applied to the Thrust Model disk velocity, where the C_T is referred to as the incident velocity factor $r\beta$. This can be seen applied to the Thrust Model disk velocity in Figure 11B. Figure 12 is the projection of the three-dimensional mapping along the β *axis* as shown in Figure 11A,B.

When mapped in three dimensions, the C_T domain in the β dimension is bounded by non-dimensionalizing C_T to the local value of $r\beta v_\infty$. This maintains the AMs' frame of reference and allows direct comparison to the Thrust Model in the same C_T range. The incident velocity at the disk, $r\beta v_\infty$, is taken to be the relevant velocity for pressure drop and C_T. Disk loading remains non-dimensionalized to v_∞ (equal to the Betz, Rankine–Froude, and AM C_T). In the Thrust Model the C_T is related to $r\beta v_\infty$ and the C_P, velocity ratio, and disk loading are related to v_∞. Therefore, the Thrust Model disk loading predictions diverge from the classical two-dimensional relationship expressed in the Betz, Rankine–Froude, and AM developments where C_T and disk loading are the same.

In Figure 12 the tube case is plotted to show the difference in disk velocity (as a ratio of free stream velocity) and C_P prediction between AMs and the Thrust Model. In the Thrust Model, when $\beta = 1$ the confined rotor displays roughly a 30% loss in output power due to its confinement. This correlates well with the loss in tube C_P shown in Figure 8. No confinement is present in the AMs.

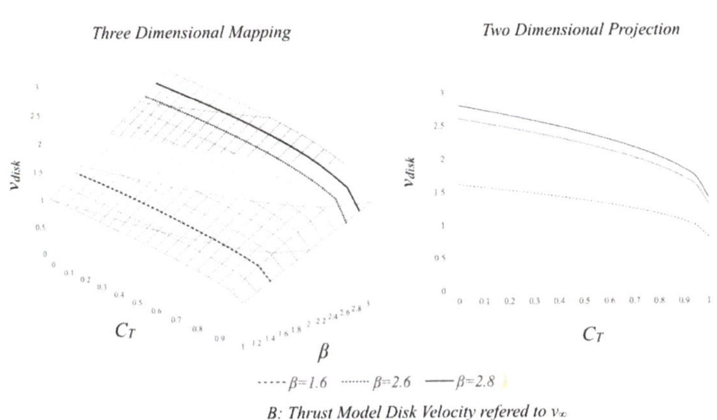

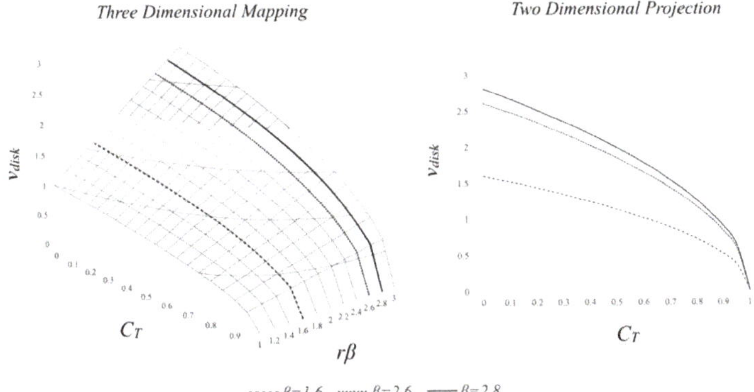

Figure 11. (**A**) General AM Disk Velocity related to C_T (**B**) Thrust Model Disk Velocity related to C_T.

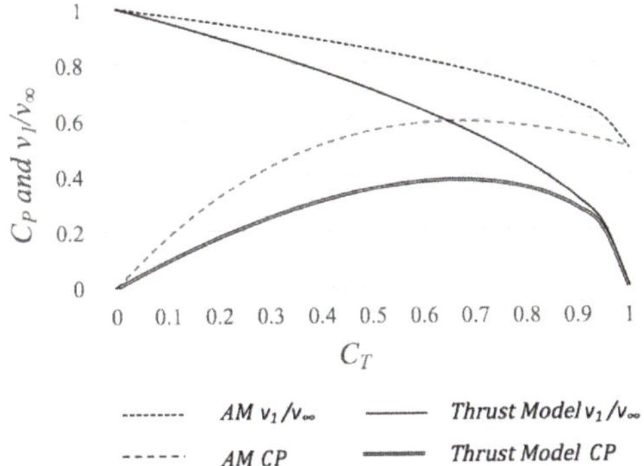

Figure 12. Thrust Model and General AM prediction of C_P and velocity ratio (v_1/v_∞) in a tube $(\beta = 1)$.

Figure 13A–C show experimental results projected into the 2D. These figures show that it is insufficient to base any validation solely on C_P without a means to qualify said C_P. Validation requires the prediction of C_P and disk velocity and pressure drop to ensure a given hypothesis is accurate in its representation of physical conditions.

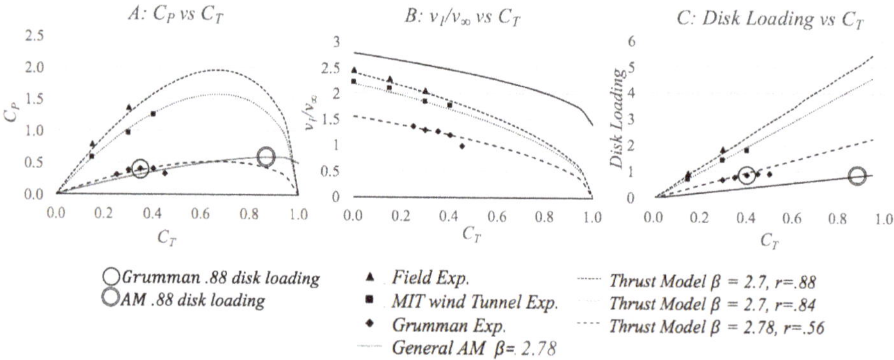

Figure 13. Thrust Model and general AM predictions compared to experimental results (**A**) C_P vs. C_T, (**B**) Velocity ratio vs. C_T, (**C**) Disk Loading vs. C_T.

Figure 13A–C also demonstrates the level of predictive error for the AM and the Thrust Model when the models are compared to the more efficient class of accelerators tested at MIT and in the field. As noted, these accelerators have significantly higher efficiencies than the Grumman DAWT. The experiments were performed with screens to represent uniform power extraction as in the Grumman experiments. The accelerator's baseline acceleration rates were 2.4 v_∞, $r = 0.88$ in the field, and 2.2v_∞, $r = 0.84$ in the wind tunnel. The effect of r on power output is non-linear. As noted above Grumman reported a baseline velocity of 1.56v_∞, $r = \sim 0.56$, for their 30-degree diffuser [10,17].

Figure 13A shows that the AMs appear to match the Grumman C_P curve reasonably well, assuming losses. Further examination of van Bussel's notes [9] on disk loading and the corollary velocity and disk loading conditions his model predicts, Figure 13B,C, shows the AM predictions do not match experiments. The 0.88 disk loading value noted as evidentiary by van Bussel is not at a concurrent performance level to the experiment. Two different C_P conditions are conflated, shown in Figure 13A, AMs yielding 0.88 disk loading at 59.3% C_P and Grumman experiments yielding 0.88 disk loading at 40% C_P [10,11]. The component variable predictions elucidate the problems with AMs. The general ideal AM model over-predicts Grumman velocity data by a factor of almost two. The inclusion of van Bussel's backpressure would increase the error in AM's predicted velocity. Jamieson and WP can account for the velocity discrepancy of the general AM by use of their variables a_0 and C_S, respectively. If van Bussel's γ was inverted, it would serve a similar function as the other AM authors correction for inefficient accelerators. Disk loading in the general AM is under-predicted by a factor of two in the Grumman case and does not vary with velocity. Jamieson's and WP's ability to correct for inefficient accelerator design would only render a further under-prediction of pressure drop. For the higher efficiency accelerators, the AMs' pressure drop error is more significant. Simultaneous large over-prediction and under-prediction illustrate the foundational flaws in the AMs.

4.2. Potential Limitations

Field testing and wind tunnel testing are the first steps toward validating the Thrust Model. More extensive testing with further instrumentation at all relevant stations described in the model is required to properly validate the Thrust Model's predictive accuracy. Such a complete data set on any accelerator is absent in the published prior art. Comparison to prior experimental data is especially useful in the case of the Grumman DAWT. The Thrust Model's accurate prediction of the Grumman experimental results, when taken in combination with the low error predictions on the new experimental data, suggests that the Thrust model is likely accurate for uniform extraction. The Thrust Model predicts other published DAWT data well. Unfortunately, this data is either restricted to C_P data or lacks the requisite velocity data when $C_T = 0$. The lack of the initial state and a complete set of component variables makes any correlation questionable given the narrow range of DAWT performance ($r = 0.4$–0.6). In all prior work other than Grumman and Phillips, the r and C_{Pdisk} variables must be estimated so the results are essentially fitted to C_P in the absence of a full set of component variables.

An interesting result of the field testing and MIT wind tunnel testing is the respective 1.4 and 1.2 system C_P. This result implies the highest known system C_P provided the data is valid. This system C_P and the component variable measurements defies the prior AMs. In terms of experimental accuracy, the tests performed in the field had complete instrumentation. Figure 14A–C shows a representative data set for the raw velocity data, pressure drop data, and power respectively for the 24.5% solidity screen test. The line in Figure 14A–C shows the best fit line with the Y-Intercept = 0. As testing was performed in the field the fluctuations are not surprising. The data set's system C_P for the 24.5% screen had a standard deviation of 0.579 based on the product of throat area, sample pressure, and sample velocity results compared to ideal power available at the intake based on ambient velocity. Despite the fluctuations, even if the -1σ are used for this data set, the 24% screen implies a minimum of 0.79 C_P. This shows that the data shows with 68% certainty that this device could exceed the Betz Limit. More testing should be performed in more steady winds or in a wind tunnel to reduce the standard deviation.

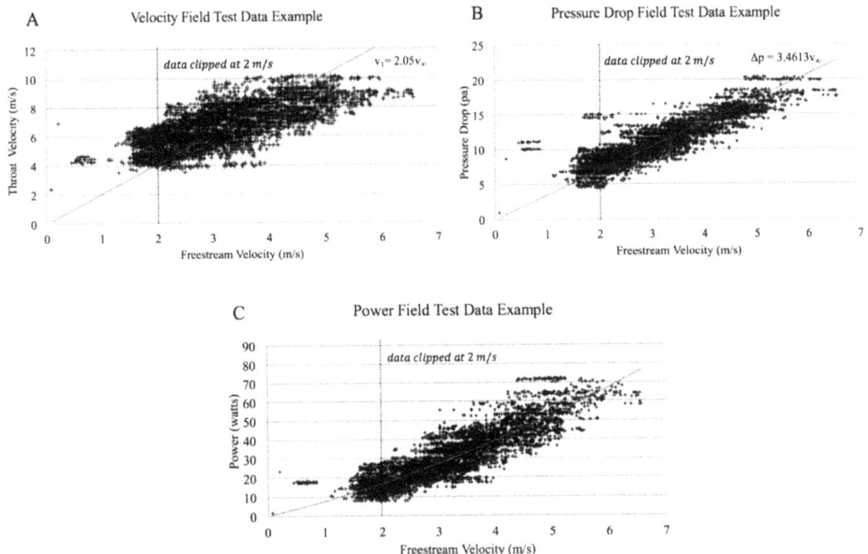

Figure 14. (**A**) shows the raw 24.5% solidity screen velocity, (**B**) shows the raw 24.5% screen pressure drop data, (**C**) shows the power as the product of throat velocity, pressure drop, and throat area.

Ambient velocities below 2 m/s led to very high multiples of power. Therefore, the data for ambient velocities below 2 m/s was clipped. The mean values of power were taken as the arithmetic means of each sample's product u_1 and Δp. These samples do include samples for which the axial orientation of the nozzle was off-center because the nozzle was manually oriented. Qualitatively, the accelerators perform well to 20° off-center, but the complete analysis is needed to determine what losses the manual orientation could have caused. For these reasons, even though there was some fluctuation in data, the results should illustrate the non-optimal performance of the ducted turbine. More testing could be performed on other efficient turbine designs to further validate the experiments.

As noted, the field testing was performed with screens at the throat to represent uniform power extraction. This method has been used by others like Grumman [17] and has been shown to provide an accurate representation of power extraction by Phillips [13]. Non-uniformity and non-ideal extraction may reduce extractable power as predicted by actuator disks and screen data.

Both the field testing and the wind tunnel testing done in the MIT wind tunnel suggest similar C_P's. The wind tunnel tests are limited because the screens were placed at the exit instead of at the throat. The resistance is experienced at the screen which covered the entirety of the exit. Therefore, the velocity is reduced before the screen. This is translated by mass flow conservation to the reduced throat velocity related to and mass rejection at the inlet. Therefore, the velocity does not have a limitation. The limitation is present for the pressure drop, as it did not occur at the throat but the exit plane, and therefore must be related to the throat theoretically. The field testing demonstrates that the pressure drop occurs at the relevant velocity, and thus the pressure drop can be translated from the exit to throat plane based on the relevant β. The experimental data sets support each other. However, wind tunnel tests with a screen at the throat would further support the findings presented herein.

The Thrust Model has limitations similar to Betz's Law. Like Betz's Law, the Thrust Model assumes incompressible flow, uniform disk loading across an infinitely thin disk, and 1-D simplifications. Like Betz's Law which should only be considered valid for freestream flow, the Thrust Model is only applicable to unpressurized confined flow, in which confinement is present in the freestream but is open on each end. A weighted coefficient could adjust either's results, but post derivation. Unifying the theories with correction coefficients serves little purpose scientifically or predictively. The Thrust Model has been shown by us to be mathematically and experimentally correct when compared to third-party tests, like Grumman and MIT, as well as testing performed by the authors. The tests show that the theory works well for actuator disk extraction inside of accelerators. Its applicability to propellers has not yet been investigated. Due to the added energy required for a propeller, it is unlikely that the Thrust Model's passive technique would directly apply to propellers. The experiments demonstrate that the pressure drop occurs at the accelerated throat velocity. Provided that the product of pressure drop and velocity at which the pressure drop occurs is the power, the Thrust Model demonstrates that significant power increases are available for ducted wind turbines over standard wind turbines.

The tested 0.72 and 1.37 system field C_P's and 0.96 and 1.25 wind tunnel C_P's are more than all other wind power in the art. This agrees well with the increase in the efficiency of the new accelerators over previously tested accelerators. To turn that result into real power, a pure uniform extractor would be required. Non-uniform power extraction will likely create added losses that would reduce the realized power multiple. There are likely different limits for different extractor device types. The Thrust Model implies that efficient high-rate acceleration could create significant power increases based on the frontal area system C_P.

5. Conclusions

The AM hypotheses are invalid because they incorrectly apply Rankine–Froude and Betz's velocity averaging methods to passive accelerators without satisfying the assumptions of Betz's Law. AMs exhibit inverse predictive error sufficient to disprove the hypotheses and contain violations of continuity. The pressure drop across a uniform resistance measured in experiments occurs at the accelerated local velocity, not at the external velocity average used by AMs. The rate of velocity reduction at the inlet occurs at a greater rate than AMs predict. Due to the use of external velocity averaging, the AMs' equations disregard the effects of the accelerator.

The Thrust Model closely predicts uniform power extraction in confined channels, matching experimental data. Further work and more complete instrumentation is required to validate the theory. The Thrust Model shows that the disk specific maximum C_P for a confined channel is 38% of the power available to the disk, significantly lower than the 59% Betz freestream limit yielded by AMs when β equals one.

C_P limits for accelerators are a function of r and β. The Thrust Model shows the C_P performance of a high-efficiency accelerator increases with increased β. This is a novel result critical to informing future R&D paths.

The qualification of specific experimental accelerators by the Thrust Model in this work shows that R&D efforts in this area have been unproductive due to the absence of benchmarking methods for accelerators. For an accelerator design to justify the research, the efficiency must be sufficient in the empty state ($C_T = 0$) to exceed the power of a freestream turbine of the same maximum frontal area. Equation (45) shows the minimum baseline efficiency for a given accelerator to justify the presence of an accelerator.

$$r > \frac{\sqrt[3]{\dfrac{0.59 A_{max}\left(v_\infty^3\right)}{0.38 A_1}}}{\beta v_\infty} \tag{45}$$

The Wind Lens [19] is the only DAWT that may have met this benchmark. When qualified, DAWTs, except the Wind Lens, have less available power than the Betz limit of a freestream turbine of their maximum area. The high-efficiency accelerator designs, tested in the field and the MIT wind tunnel, significantly surpass all DAWTs as shown by Figures 13, A4 and A7. (See Appendix A).

The result of having a system C_P greater than one is of interest. With a system C_P greater than one, the system would extract more energy than the kinetic energy in the streamtube of the maximum area of the accelerator projected forward. This phenomenon seems to contradict the assumption in ideal incompressible external fluid dynamics, that the only power available for extraction is the kinetic energy in the flow of the maximum frontal area projected forward. The mathematics and laws of continuity contradict the validity of that assumption. While not mathematically supported to date this prior assumption can, but should not be ignored due to its pervasiveness in the art. Several possible explanations could resolve these contradictions.

The accelerator extractor pair could draw kinetic energy from an area greater than its frontal area. This could be likened to an object, like a flat plate, which has a coefficient of drag greater than one. This larger area would be related to Equation (27) above. This larger area could be referenced to mass flow or pressure or a combination of the two. In this way the flow is non-ideal.

The second explanation in field testing could be related to local pressure gradients, causing gusts, lulls, and transient effects. With field testing, fluctuations in wind speed are expected and for this reason, many data sets were taken. Due to the extensive testing time, the local pressure gradients and velocity variations should not have significantly affected results. However, more testing needs to be done to confirm these results.

Due to the combination of confined (internal flow) and freestream (external flow) phenomena, the inherent assumptions based on freestream kinetic energy may be incorrect. The local pressure variations caused by the accelerator suggest that kinetic energy assumptions assigned for steady-state velocity driven, incompressible, and ideal flow may not be valid.

The standard assumption that the power is limited to the ideal kinetic energy of the frontal area could be true. However, this assumption is what the AMs are predicated upon and leads to AMs' discontinuities. Furthermore, this would require that the field, MIT, and Grumman testing was incorrect. The use of actuator disk theory inside of a duct could be misapplied, however, this is also unlikely due to Phillips and Grumman's research. For these reasons, the thrust model predictions have been confirmed with three experimental data sets. For two of these, the authors were not involved in the experiments, MIT and Grumman. The Thrust Model should be applied to more experimental data. However, complete data sets, which include uniform extraction at the throat and baseline accelerator performance, are unavailable in the art.

The flow could cease to be incompressible at the points of extraction, but this is unlikely due to the low speeds. This analysis assumes incompressible flow; therefore, density is assumed to be constant.

More research is required to determine if there is another cause for these apparent contradictions. Further work is necessary to determine accelerators' optimal performance for non-uniform power extraction, the optimal form of the technology, and the deployed power density of actual systems which is dependent on the specific axial induction to C_P ratio, wake recovery length, and machine separation. Preliminarily, the Thrust Model's C_P predictions and the new experimental evidence show that passive acceleration can produce large increases in uniformly extracted fluid power. Acceleration technology applied to fluid resources could radically expand the technical potential and siting opportunities for peaking wind energy and baseload freestream hydropower. The findings in this paper show that acceleration could constitute a crucial and, as yet untapped, enabling technology for large scale fossil fuel offsets to combat climate change.

6. Patents

The research is partially a result of work for the U.S. Army. The following awarded patents are related to the research above:

1. PAT. NO.10,408,189; Efficient systems and methods for the construction and operation of mobile wind power platforms.
2. PAT. NO.9,709,028; Efficient systems and methods for the construction and operation of mobile wind power platforms.
3. PAT. NO.8,937,399; Efficient systems and methods for the construction and operation of mobile wind power platforms.
4. PAT. NO.8,598,730; Modular array wind energy nozzles with truncated catenoidal curvature to facilitate airflow.
5. PAT. NO.8,482,146; Efficient systems and methods for the construction and operation of accelerating machines.
6. PAT. NO.8,395,276; Modular array wind energy nozzle with increased throughput.
7. PAT. NO.8,178,990; Wind power nozzle with increased throughput.
8. PAT. NO.8,089,173; Wind power nozzle with optimized intake length.
9. PAT. NO.7,804,186; Modular array fluid flow energy conversion facility

Author Contributions: Conceptualization, R.F. and B.K.; methodology, R.F. and B.K.; software, R.F. and B.K.; validation, R.F., B.K., and S.P.; formal analysis, R.F. and B.K.; investigation, R.F., B.K., and S.P.; resources, R.F., B.K., and S.P.; data curation, R.F. and B.K.; writing—original draft preparation, R.F. and S.P.; writing—review and editing, S.P.; visualization, S.P.; supervision, R.F.; project administration, R.F.; funding acquisition, R.F. All authors have read and agreed to the published version of the manuscript.

Funding: This research was partially funded by the US Army under contract number W911QY-3-C-0054.

Acknowledgments: The authors would like to acknowledge the work of S. Widnall and S. Gomez for their independent research at the Wright Bros on the accelerator arrays designed by the authors. Thanks to S. Widnall for the review of the prior art analysis and Thrust model derivation.

Conflicts of Interest: The authors declare no conflict of interest.

Nomenclature

β	Area Ratio
Δp	Pressure drop across extractor
ρ	Density
a	Axial Induction
A_1	Throat area in front of extractor/Area of extractor
A_2	Throat area behind extractor
A_3	Diffuser exit area
A_{disk}	Area of disk
$A_{effective}$	Effective area of the device presented to the freestream flow
A_{in}	Inlet area
A_{max}	Maximum frontal Area of Diffuser
C_P	Coefficient of Power
$C_{Psystem}$	Coefficient of Power of max area
C_{Pdisk}	Coefficient of Power of disk
C_S	Shroud Force Coefficient used by Werle and Presz
C_T	Coefficient of Thrust
C_{Tv1}	Coefficient of Thrust non-dimensionalized by disk velocity
F_{1o}	Force available at the throat without extraction
$F_{1extracted}$	Force extracted at throat
$F_{1remaining}$	Force not extracted
F_s	Shroud force variable used by Werle and Presz
p_∞	Freestream static pressure
p_1	Throat static pressure in front of extractor
p_2	Throat static pressure behind extractor
p_3	Diffuser exit static pressure
p_{in}	Static pressure at inlet
p_{tot}	Total pressure
p_{wake}	Wake static pressure
P	Power
P_{Betz}	Power predicted by Betz's law
P_{AM}	Power predicted by Averaging Methods
$P_{accelerator}$	Power available for an accelerator
$P_{available}$	Power available
$P_{available,o}$	Initial Power available
$P_{freestream}$	Power available for a free stream turbine
q_∞	Freestream dynamic pressure
q_{wake}	Wake dynamic pressure
r	Accelerator efficiency
$r_{Grumman}$	Accelerator efficiency of the Grumman DAWT
TSR	Tip Speed Ratio
v_∞	Freestream velocity
v_1	Throat velocity in front of extractor, 'Disk Velocity'
v_{1o}	Initial Throat Velocity with no extractor present
v_2	Throat velocity behind extractor
v_3	Diffuser exit velocity
v_{disk}	Velocity at disk
v_{in}	Inlet velocity
v_{wake}	Wake velocity

Appendix A.

Appendix A.1. Specifics of AM Methods and Other Theories

In terms of the fundamental averaging equations, there is little difference between the AMs. The axial induction forms of the Betz averaging equations found in the AMs are Equations (A1) and (A2) below. Regardless of the variables used by the prior hypotheses, the respective equation systems simplify the general AM system described above. The AMs require an accelerator to have an efficiency greater than one to exceed the Betz Limit of the accelerator's frontal area. Of the previous theorists, only van Bussel addresses the maximum area and presents backpressure as a means to exceed ideal accelerator efficiency. Jamieson and WP do not estimate system efficiency, performance, or potential. Their analyses are restricted to the rotor area and characterize performance independent of the maximum area correction.

$$v_{disk} = (1-a)v_\infty \tag{A1}$$

$$v_{wake} = (1-2a)v_\infty \tag{A2}$$

Appendix A.1.1. AM Method 1: van Bussel

Van Bussel [10] derives velocity conditions in the accelerator in his Equations (2), (5), and (7)–(10). This velocity development relates the throat velocity to the exit velocity based on the axial induction. The pressure drop across a turbine in an accelerator is given by van Bussel's Equation (13) or equation A3. The pressure drop across p_2 and p_1 does not use the flow variables in the accelerator, but is derived from the axial induction, and thereby freestream and wake velocities.

$$p_1 = p_0 + [1 - \beta^2\gamma^2(1-a)^2]\frac{1}{2}\rho v_\infty^2 \tag{A3}$$

$$p_2 = p_0 + [(1-2a)^2 - \beta^2\gamma^2(1-a)^2]\frac{1}{2}\rho v_\infty^2 \tag{A4}$$

$$p_2 - p_1 = 4a(1-a)\frac{1}{2}\rho v_\infty^2 \tag{A5}$$

The axial induction is calculated based on the averaging of freestream and wake velocities, like Betz's Law. Van Bussel explicitly demonstrates his use of averaging in his specification of axial induction, "the axial induction factor a is defined at the exit of the diffuser. Just as in an ordinary wind turbine momentum theory this induction is half the induction factor found in the far wake behind the DAWT." [10]

This averaging is inherent to the derivation of the axial induction form of the Betz disk and wake velocity equations. The presence of these variables shows that van Bussel's method has the general AM errors inherent to the averaging. Additionally, the application of β to the induction terms is improper as induction should be calculated by applying β to velocity within the squared term as noted in the general AM discussion. Applying β in van Bussel's way serves to eliminate it from the axial induction equations as is noted in the analysis of the explicit equations described in the AM section. Van Bussel's Equations (11) and (12), or Equations (A3) and (A4), reduce to the general AM pressure drop or Betz equation without any nozzle variables or confinement of the rotor.

This yields van Bussel's C_P equations. Van Bussel's Equations (14) and (15), or Equations (A6) and (A7), relate C_P to the throat and exit, respectively, wherein the nozzle has been effectively eliminated from the pressure drop component. This yields the AM general result of the maximum limit being the Betz limit of the exit, as shown by van Bussel's Equation (15) or equation A7.

$$C_{P,rotor} = \beta\gamma 4a(1-a)^2 \tag{A6}$$

$$C_{P,exit} = \gamma 4a(1-a)^2 \tag{A7}$$

As can be seen in van Bussel's Equation (15) or equation (A7), van Bussel's development states that an accelerator can only exceed the equivalent freestream turbine if γ is greater than one. This use of backpressure is characteristic of active systems wherein choking the throat of a nozzle is the objective. The increased backpressure of DAWTs has not led to improved performance because increasing fluid velocity through backpressure requires the addition of energy. High backpressure is a sign of poor accelerative efficiency in passive systems and should not be an object of accelerator design.

Further, van Bussel conflates ideal and non-ideal states in his proposed validation of his momentum theory. He shows that his theory predicts the maximum power at 8/9th's disk loading and concludes that the momentum theory is validated by the 0.88 experimental disk loading of the Grumman device at the maximum tested output. This conflates the optimum condition which van Bussel asserts is 16/27th's of the maximum area with the non-optimal performance of the Grumman device. In Grumman experiments, the disk loading value of 0.88 was reported at 40% C_P related to the maximum area, not at 59% C_P which is van Bussel's result.

Appendix A.1.2. AM Method 2: Jamieson

Jamieson [11] states that the maximum coefficient of power for a shrouded rotor is the product of the Betz Limit times $(1 - a_0)$. His variable a_0, in his Equation (15) or Equation (A8), is the axial induction at the extraction plane under no extraction. For an ideal accelerator, Jamieson states, $\beta = 1 - a_0$. Therefore, Jamieson predicts that the power limit of an ideal accelerator is the Betz Limit of its maximum frontal area. His specific application of averaging in his derivation is found in his Equation (10) or equation (A9), which is his adaptation of the Betz C_T equation. Jamieson uses an axial induction reference plane, b, rather than the throat axial induction reference plane, a.

$$C_{pm} = \frac{16}{27}(1 - a_0), \text{ where } C_{pm} \text{ is maximum } C_P \tag{A8}$$

$$C_T = 4b(1 - b) = \frac{2\Delta p}{\rho V^2}, \text{ where } V = v_\infty \tag{A9}$$

Jamieson states that "Denoting axial induction at the reference plane as b, then velocity in the far wake is $V(1 - 2b)$, a result first determined by R E Rankine–Froude." Regardless of whether the typical a plane or the b plane is used, the fundamental error of applying averaging remains evident in the $1 - b$ and $1 - 2b$ terms.

Jamieson claims to have developed a general theory for open flow and confined rotors by introducing the b reference plane, but his hypothesis exhibits the same techniques, errors, and violations as the general AM. Applying the $\beta = 1$ condition shows that Jamieson's theory does not constitute a general theory for confined flow. In the $\beta = 1$ case Jamieson's a and b planes are coincident, and he asserts that the confined disk is in the same condition as an open flow rotor. Therefore, there is no confinement of the actuator disk evident.

Jamieson provides his a_0 term, which reduces the mass flow due to accelerator inefficiency. Additionally, Jamieson's control volume differs from Figures 7 and 9 above since his equations are developed with a reference plane. The reference plane designation asserts that the pressure drop occurs over an area equal to the initial mass flow rate through the accelerator at the freestream velocity. With non-ideal acceleration, Jamieson predicts that there is more thrust on the disk as compared to van Bussel, who asserts that the pressure drop occurs at the rotor size at the freestream velocity.

Appendix A.1.3. AM Method 3: Werle and Presz (WP)

WP [12] state that their equation system yields the freestream turbine when the shroud force is reduced to zero, "Note that the unducted wind/water turbine case is recovered using $C_S = 0$". In WP's framework, a $C_S = 0$ should represent a tube, not a freestream actuator disk. WP's C_s variable is the inverse of Jameison's a_0 variable. WP's Equation (2) and (3a), or (A10) and (A11), show the velocity averaging used for the calculation of disk velocity. WP's Equation (4), or Equation (A12), shows the equivalency between C_S and Jamieson's a_0 referring to Jamieson's Equation (14) [11]. Where possible, the nomenclature is converted from WP's nomenclature to the nomenclature of this paper.

$$F_s = \frac{1}{2}\left[\rho A_{disk}\left(v_{wake}^2 - v_\infty^2\right)\right]C_S \tag{A10}$$

$$v_p = \frac{1}{2}(1 + C_s)(v_{wake} + v_\infty) \tag{A11}$$

$$C_{Pmax} = \frac{16}{27}(1 + C_S) \tag{A12}$$

WP's work exhibits the same fundamental mathematical structure, assumptions, and errors as Jamieson and the general AM.

Appendix A.1.4. Non-AM Method: Lawn

Lawn [24] developed an analytical model to analyze extractors inside of accelerators. Lawn defines the resistance coefficient, K, in his Equation (1), and defines the coefficient of power in his Equations (6) and (7). Accelerator efficiency and turbine efficiency is accounted for in his model development. Unlike the AMs, Lawn does not claim to present a theoretical model that constrains the power and flow effects of an accelerator. Instead, Lawn derives his velocity and pressure drop data from experiments and references. Since Lawn does not use velocity averaging to determine flow conditions, he can accurately determine results if correct flow variables are input. For example, Lawn predicts that an extractor within a tube does not behave like a freestream turbine but is rather restrained to lower power. Lawn's model is limited since it is an analytical model and it cannot provide a means to evaluate designs without experimental evidence.

Appendix A.2. Methods for New Experiments: Third-Party High-Efficiency Array testing in MIT Wind Tunnel and High-Efficiency Accelerator Field Testing

The AMs and the proposed Thrust Model are compared to several experiments: the Grumman DAWT experiments, a high-efficiency accelerator array tested in the MIT wind tunnel, and field testing of a high-efficiency accelerator. The Grumman device with a screen in place is pictured in Section 1 and the specifics of the Grumman experiment can be found in reference [17]. Below are the procedures for MIT and the authors' experiments.

Appendix A.2.1. Accelerator Array Tested in MIT Wind Tunnel

A 1:4 scale accelerator array was tested in the MIT Wright Brothers Wind Tunnel [25], as shown by Figure A1A. The design for the array was provided by the authors. The tests were performed by S. Gomez. The array consisted of a 3×5 array of accelerators where $\beta = 2.6$. Screens were attached to the exit of the array. An X-Y traverse was used to measure the pressure across the exit plane of the array. The screens were placed at the exit to measure the complete radial exit flow field with the X-Y traverse which would not have been possible at the individual throats.

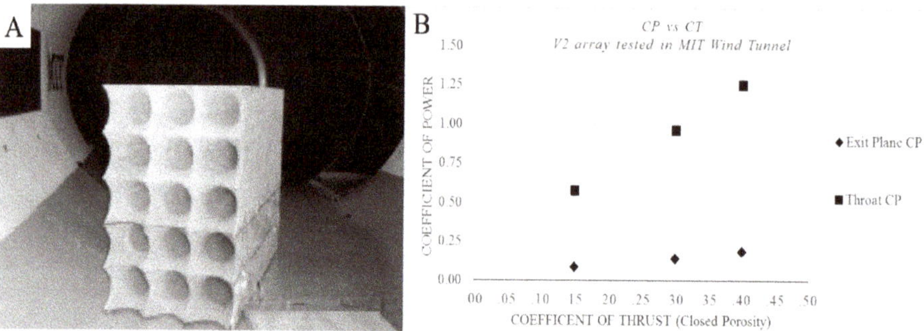

Figure A1. (**A**) Stagnation pressure measurements were taken for screens with open porosities of 60.4%, 74.6%, 87.8%, and 100% (no screen). The same screens were tested at the exit of the array and in free-standing screen tests to isolate C_P correlation to screen open porosity. The exit screen will produce resistance in accordance with the given C_T value and that velocity reduction will be experienced at the throat and intake related to local β. The tested C_P at the exit is related by mass flow conservation to a throat C_P by the relation of $(r\beta)^2$. (**B**) depicts the experimental exit plane C_P and the correlating throat plane C_P. The C_P is non-dimensionalized based on freestream velocity and maximum frontal area. Figure A2 depicts the mass flow rate for the array at a different closed percentage.

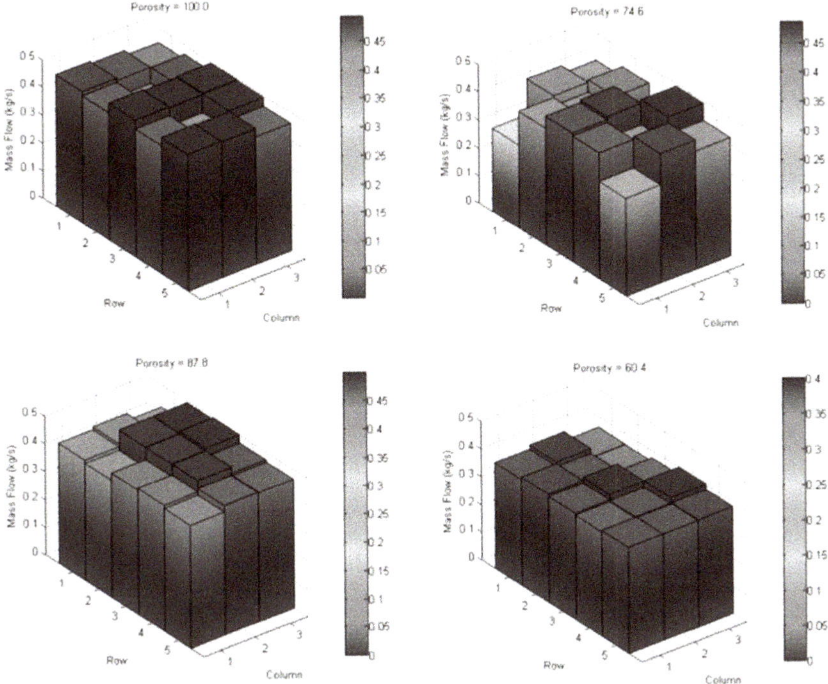

Figure A2. MIT Wind Tunnel Results: Mass flow in each accelerator of the array for 100%, 87.8%, 74.6%, and 60.4% porous arrays [25].

There are limitations to the results of the experiments presented by Gomez. The first is that, as shown in Figure A2, there are inconsistencies in mass flow between the different accelerators, which could be due to turbulence. Further testing should be performed to attempt to gather more uniform results, especially for the 74.6% porosity case. The second limitation is due to testing constraints to use the X-Y traverse, the screen was tested at the exit of the accelerator instead of at the throat. This required further analysis to relate the exit plane performance to the throat plane.

Appendix A.2.2. Field Experiment for a U.S. Army Phase I SBIR

A single full-scale wind acceleration module was designed and fabricated for testing in the field [26]. The test nozzle dimensions were 1 m × 1 m × 5 m. The nozzle was attached to a bearing assembly and mounted on a ~1.5 m tall quadpod. Figure A3A shows the inlet, center body assembly, exit assembly, and bearing and elevation assembly. Velocity measurements without screens were taken to characterize the baseline acceleration performance of the prototype, *r*. The test setup is shown in Figure A3B. Rotor anemometer measurements were taken to confirm the main pressure measurement system velocity data. Hotwire anemometer spot checks were performed throughout testing with Sper hotwire anemometers.

Velocity and power testing were performed using screens at the throat to simulate a uniform actuator disk. Pressure and anemometer measurements were used to capture results. The pressure system was comprised of three sets of Prandtl tubes to measure velocity at the throat, with and without screens, and the pressure drop across the screens to estimate power output. Pressure measurements were taken by differential pressure transmitters. Ambient wind speed and direction data were taken with 2 co-mounted SKADA cup anemometers, as shown by Figure A3C. All pressure and anemometer signals were recorded to a datalogger to insure the simultaneous measurement of the data. Figure A3D shows the instrumentation panel on the side of the prototype. Prandtl tubes were arranged in pairs. The forward tube measured stagnation and static pressure before the throat/screen. The rearward tube measured static pressure after the screen. In the field, the rearward tubes were oriented downstream. Prandtl static ports were tested in a wind tunnel and the field with forwarding and rearward orientation and no difference was found in the static pressure measurement. Rearward tubes did not require stagnation measurement. Figure A3E,F depicts the instrumentation at the throat.

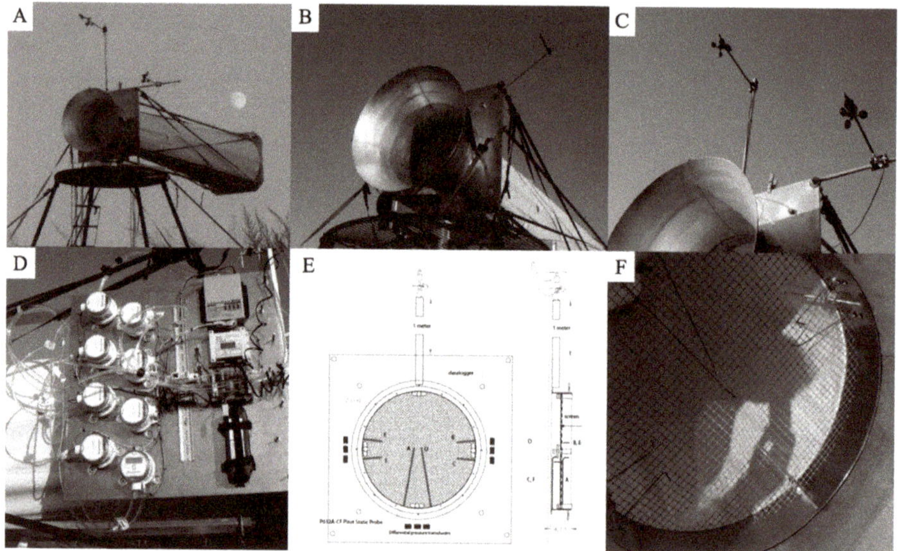

Figure A3. (**A**) Accelerator Field Experiment, (**B**) Rotor anemometer velocity measurement, (**C**) Anemometers 1m above and to the right of the accelerator, (**D**) Instrumentation panel on side of prototype, (**E**) Diagram of Prandtl tube arrangement in the accelerator, (**F**) Prandtl Tubes installed with a 24% solidity screen in the throat of Accelerator.

Screens used in testing were 15.2% and 24.5% closed porosity. Previous testing in MIT's Wright Brothers wind tunnel, referenced above, indicated that these screens represented turbine C_P's of roughly 15% and 28%, respectively.

Figure A4 below depicts the experimental multiple of the velocity, pressure drop, and ambient power for 0%, 15%, and 24% solidity screens. On average each data set was one hour long, and data was collected for over 20 h. Instantaneous samples were taken every half second. Each sample's power was determined by the product of v_1, A_1, and Δp as shown by Equation (A13). The average of all samples above a v_∞ of 2 m/s is used.

$$P_{screen} = \Delta p_1 A_1 v_1 \tag{A13}$$

Velocity ratio is the achieved throat velocity, $r\beta$, related to ambient velocity. The pressure drop ratio is taken as the square root of the ratio between the measured pressure drop and the ideal Betz pressure drop calculated at the ambient velocity. C_P ratio is the ratio between the extracted power at the accelerator's throat and the total fluid power available at the ambient velocity at the accelerator's maximum area.

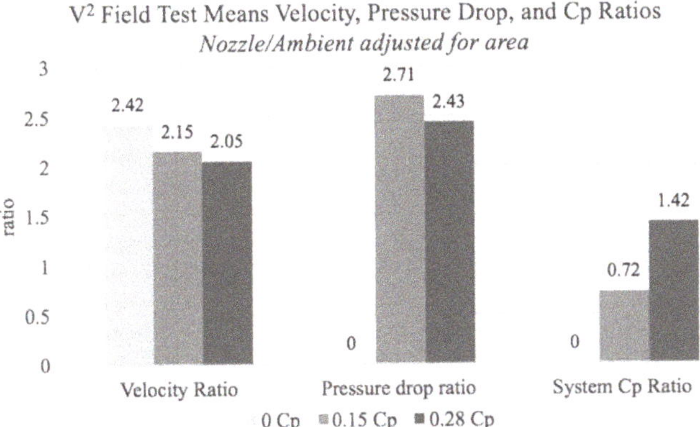

Figure A4. Mean Test data ratio to freestream turbine variables at 0, 0.15, and 0.28 rotor C_P.

Appendix A.3. Experimental Results Shown by Non-Dimensionalizing Thrust with Disk Velocity

To further illustrate the predictive errors with the AMs and accuracy of the Thrust Model, the C_T can instead be non-dimensionalized with v_1. A new variable, C_{Tv1}, is created by non-dimensionalizing C_T with v_1 instead of v_{10} (Thrust Model) or with v_∞ (AMs). C_{Tv1} provides an alternate metric to compare all models non-dimensionalized with the same station velocity in a way that is commonly used for the internal flow. Figure A5 shows the disk velocity and C_P predictions for the Thrust Model and the general AM. As the general AM is only able to produce results up to $a = 0.5$, the maximum C_{Tv1} is 4 (as the disk velocity is 0.5, so $C_{Tv1} = C_T/0.5^2 = 4$). In contrast, the Thrust Model can analyze results until disk velocity is zero at which point C_{Tv1} is infinite. This image also shows that the general AM predicts the maximum power to be the Betz Limit of the maximum area instead of 38% C_P as predicted by the Thrust Model.

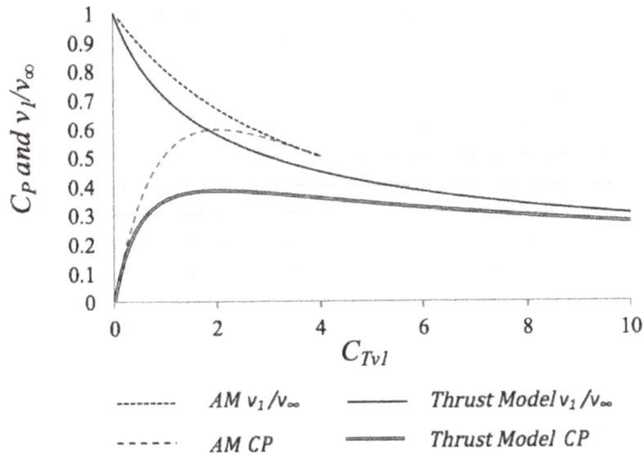

Figure A5. General AM and Thrust Model predictions of v1 and C_P vs. C_{Tv1} in a tube ($\beta = 1$).

Figure A5A–C depicts the disk velocity, C_P, and disk loading for the Grumman experiment, the general AM prediction, the Thrust model prediction, Jamieson's prediction, van Bussel's prediction, and van Bussel's prediction if his γ variable was inverted. WP was not plotted due to its similarity to Jamieson's model. To note, when the Thrust Model is non-dimensionalized with disk velocity, as the disk velocity approaches zero, the C_{Tv1} approaches infinity. In contrast, models like van Bussel are only able to analyze results to ~0.5 axial induction entering the accelerator. For this reason, van Bussel's model and the general AM do not have a C_{Tv1} values that correlate with the Grumman data. This depicts how van Bussel's model and the general AM over-predict disk velocity and under-predict corollary pressure drop. If van Bussel's γ was inverted he would be able to model disk velocity more accurately but would still over-predict. In Figure 13 above the AM C_P predictions as a function of C_T

are coincidently close to the Grumman results. When plotted based on C_{Tv1}, the AM predictions are no longer close to the Grumman results. This method removes the possibility for conflation of C_P when component variables are incorrectly predicted.

Jamieson's a_0 variable reduces the initial mass flow through the device and as a result, can predict disk velocity vs. C_{Tv1} more accurately than van Bussel and the general AM. Despite Jamieson's ability to calculate disk velocity, he underpredicts the pressure drop, except for the highest C_{Tv1} data point, since his model assigns pressure drop at the freestream velocity. As a result, Jamieson underpredicts C_P. Figure A6 is plotted using his Equations (A1) and (A2).

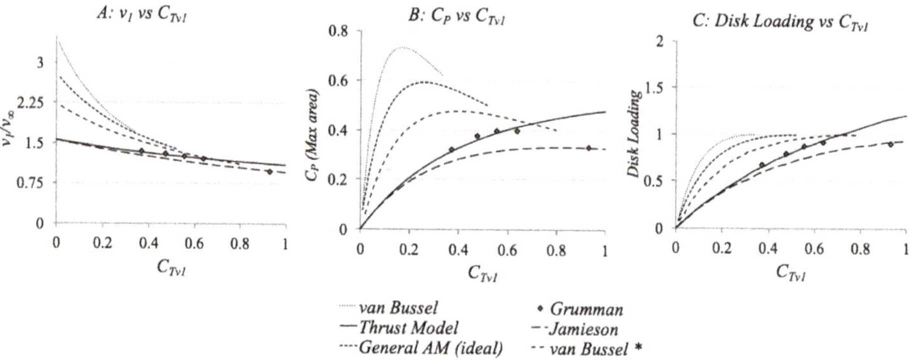

Figure A6. Grumman Test Results compared to theoretical models (**A**) v1 vs. C_{Tv1} (**B**) C_P vs. C_{Tv1} (**C**) Disk Loading vs. C_{Tv1} (* Denotes van Bussel's model with inverted back pressure which more accurately predicts results).

In contrast, the Thrust Model accurately predicts the Grumman data except for the highest C_{Tv1} data point. Figure A6 demonstrates the accuracy of the Thrust Model, especially when compared to the AMs. The Thrust Model can accurately predict disk velocity, disk loading, and power.

Figure A7 illustrates the field-testing results compared to the general AM, Thrust Model, and Jamieson predictions as a function of C_{Tv1}. As shown above, in Figure 13, van Bussel's backpressure leads to over prediction of velocity so it is not included, since the general AM will be closer to real results. As Figure A7 depicts, the general AM and Jamieson underpredict the disk loading and C_P for the field test data. The AM's apparent underprediction of v_1 vs. C_{Tv1} is caused by the underprediction of pressure drop, which shrinks the AM's C_{Tv1} domain. The Thrust Model accurately predicts (within 12%) the velocity, disk loading, and C_P of the field tests. The Thrust Model predicts the lower porosity screen more accurately than the higher porosity screen, where it over predicts by 8% and under-predicts by 12%, respectively.

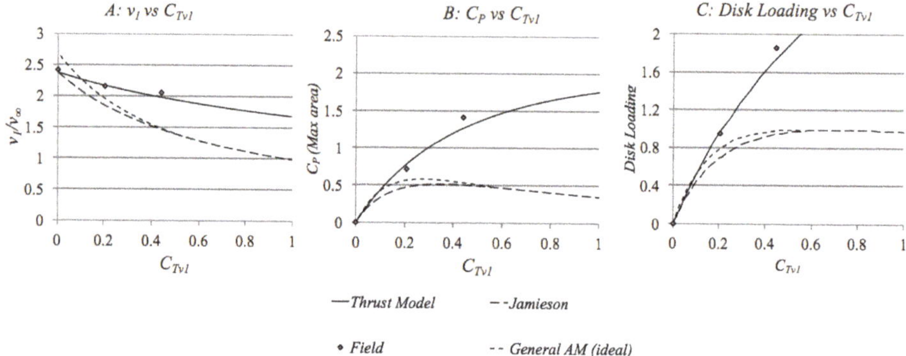

Figure A7. General AM, Thrust Model, and Jamieson's model compared to Field Test Results for (**A**) v_1 vs. C_{Tv1} (**B**) C_P vs. C_{Tv1} (**C**) Disk Loading vs. C_{Tv1}.

The Thrust Model's predictions correlate well to all experimental results, prior art data as well as the new experimental data. The largest C_P error is 12% for the 24.5% screen field tests. The Thrust Model yields a disk C_P of 0.28 and a disk-loading coefficient of 0.84 for $r = 0.56$. Grumman disk-loading was 0.88. System C_P for the Thrust Model prediction was 38%, compared to Grumman at 40%. The throat velocity for the Thrust Model was $1.2v_\infty$ compared to Grumman at $1.2v_\infty$ [17]. The accuracy of the Thrust Model's prediction confirms that the pressure drop occurs at the accelerated velocity not at the average of the external velocities. C_P of 0.28 for a free stream disk would correspond to a disk loading of ~0.3 which correlates to the 2.78 β value for the Grumman accelerator ($0.3 \times 2.78 = 0.84$). This confirms that throat disk loading is related to intake disk loading by β and pressure drop is related by β^2 as shown in Equation (23) above.

References

1. Carlos, J.; Martin, A. Key Performance Indicators Wind Farm Availability TIME Vs ENERGY. In *Analysis of Operating Wind Farms*; EWEA Technology Workshops: Lyon, France, 2012.
2. Smil, V. Power Density Primer: Understanding the Spatial Dimension of the Unfolding Transition to Renewable Electricity Generation; 8 May 2010. Available online: http://vaclavsmil.com/wp-content/uploads/docs/smil-article-power-density-primer.pdf (accessed on 24 June 2020).
3. Sansoucy, G. In-City Renewable Resource Executive Summary for the Theoretical, Technical, and Economic Potential for Renewable Energy Resource Development in the City and County of San Francisco as part of the CCA Program; CCA Contract No.: CS No.: CS-920R-A; CCA 2009. Available online: https://sfgov.org/lafco/sites/default/files/FileCenter/Documents/19063-GES_Draft_Report_Tasks1-2.pdf (accessed on 24 June 2020).
4. Schmalensee, R. *The Future of the Us Electric Grid*; MIT Press: Cambridge, MA, USA, 2016; pp. 73–79.
5. Lopez, A.; Roberts, B.; Heimiller, D.; Blair, N.; Porro, G.U.S. *Renewable Energy Technical Potentials: A GIS-Based Analysis*; Technical ReportNREL/TP-6A20-51946; Office of Scientific and Technical Information (OSTI): Oak Ridge, TN, USA, 2012.
6. California Power Plant Database.xls. Available online: https://www.eia.gov/electricity/data/eia923/ (accessed on 28 June 2020).
7. *Power Trends 2015: Rightsizing the Grid*; NYISO: Rensselaer, NY, USA, 2015.
8. Gurney, J.; Company, B.P. BP Statistical Review of World Energy. *J. Policy Anal. Manag.* **1985**, *4*, 283. [CrossRef]
9. Betz, A. *Introduction to the Theory of Flow Machines*; Elsevier BV: Amsterdam, The Netherlands, 1966.
10. Van Bussel, G.J.W. The science of making more torque from wind: Diffuser experiments and theory revisited. *J. Phys. Conf. Ser.* **2007**, *75*. [CrossRef]
11. Jamieson, P. Beating Betz—Energy Extraction Limits in a Uniform Flow Field. In Proceedings of the European Wind Energy Conference, Brussels, Belgium, 1 January 2008; Available online: http://www.katru.com.au/External%20Recognition/13_EWEC2008fullpaper.pdf (accessed on 5 June 2020).
12. Werle, M.J.; Presz, W.M. Ducted Wind/Water Turbines and Propellers Revisited. *J. Propuls. Power* **2008**, *24*, 1146–1150. [CrossRef]
13. Phillips, D. An Investigation of Diffuser Augmented Wind Turbine Design. Ph.D. Thesis, University of Auckland, Auckland, New Zealand, 2003.
14. Kline, S.J.; Abbott, D.E.; Fox, R.W. Optimum Design of Straight-Walled Diffusers. *J. Basic Eng.* **1959**, *81*, 321–329. [CrossRef]
15. Knight, B.; Freda, R.; Young, Y.L.; Maki, K. Coupling Numerical Methods and Analytical Models for Ducted Turbines to Evaluate Designs. *J. Mar. Sci. Eng.* **2018**, *6*, 43. [CrossRef]
16. Rankine-Froude, R.E. On the part played in propulsion by differences of fluid pressure. *Trans. Inst. Nav. Archit.* **1889**, *30*, 390–405.
17. Oman, R.A.; Foreman, K.M.; Gilbert, B.L. *Investigation of Diffuser-Augmented Wind Turbines: Part 2*; Technical Report DOE/SERI #Contract No. EY-76-C-02-2616; Research Department Grumman Aerospace Corporation: Bethpage NY, USA, 1977.
18. Concentrating Windsystems—Sense or Nonsense. Available online: http://www.heiner-doerner-windenergie.de/diffuser.html (accessed on 3 May 2016).
19. Gipe, P. Ogin Installs Ducted Turbines in the San Gorgonio Pass" & "FloDesign-Ogin Some Brief Comments (CP claims). Available online: http://www.wind-works.org/cms/index.php?id=665 (accessed on 3 June 2020).

20. Ohya, Y.; Karasudani, T. A Shrouded Wind Turbine Generating High Output Power with Wind-lens Technology. *Energies* **2010**, *3*, 634–649. [CrossRef]
21. Igra, O. Shrouds for Aerogenerators. *AIAA J.* **1976**, *14*, 1481–1483. [CrossRef]
22. Reid, E.G. Performance Characteristics of Plane-Wall Two-Dimensional Diffuser. In *NACA Technical Note 2888*; Nobel Press: Berlin, Germany, 2012.
23. Bussel, G. An Assessment of the Performance of Diffuser Augmented Wind Turbines (DAWT's). In Proceedings of the Third ASME/JSME Joint Fluids Engineering Conference, San Francisco, CA, USA, 18–23 July 1999.
24. Lawn, C.J. Optimization of the power output from ducted turbines. *Proc. Inst. Mech. Eng. Part A J. Power Energy* **2003**, *217*, 107–117. [CrossRef]
25. Gomez, S. Performance of a Ducted Wind Turbine Array Final Report. In *Massachusetts Institute of Technology Report*; MIT Press: Cambridge, MA, USA, 2011; p. 16.622.
26. Freda, R. *SBIR Topic Number A12-051—Phase I Final Report*; United States Army: Boston, MA, USA, 2013.

Article

Numerical Simulation of the Solid Particle Sedimentation and Bed Formation Behaviors Using a Hybrid Method

Md Abdur Rob Sheikh [1], Xiaoxing Liu [1], Tatsuya Matsumoto [1], Koji Morita [1,*], Liancheng Guo [2], Tohru Suzuki [3] and Kenji Kamiyama [3]

[1] Department of Applied Quantum Physics & Nuclear Engineering, Kyushu University, Fukuoka 819-0395, Japan; rob_baec@yahoo.com (M.A.R.S.); liuxx85@mail.sysu.edu.cn (X.L.); matumoto@nucl.kyushu-u.ac.jp (T.M.)
[2] Institut für Kern- und Energietechnik, Karlsruher Institut für Technologie, 76344 Eggenstein-Leopoldshafen, Germany; Liancheng.Guo@grs.de
[3] Fast Reactor Cycle System R&D Center, Japan Atomic Energy Agency, Ibaraki 311-1393, Japan; torusuzu@tcu.ac.jp (T.S.); kamiyama.kenji@jaea.go.jp (K.K.)
*   Correspondence: morita@nucl.kyushu-u.ac.jp; Tel.: +81-92-802-3498

Received: 1 August 2020; Accepted: 21 September 2020; Published: 24 September 2020

Abstract: In the safety analysis of sodium-cooled fast reactors, numerical simulations of various thermal-hydraulic phenomena with multicomponent and multiphase flows in core disruptive accidents (CDAs) are regarded as particularly difficult. In the material relocation phase of CDAs, core debris settle down on a core support structure and/or an in-vessel retention device and form a debris bed. The bed's shape is crucial for the subsequent relocation of the molten core and heat removal capability as well as re-criticality. In this study, a hybrid numerical simulation method, coupling the multi-fluid model of the three-dimensional fast reactor safety analysis code SIMMER-IV with the discrete element method (DEM), was applied to analyze the sedimentation and bed formation behaviors of core debris. Three-dimensional simulations were performed and compared with results obtained in a series of particle sedimentation experiments. The present simulation predicts the sedimentation behavior of mixed particles with different properties as well as homogeneous particles. The simulation results on bed shapes and particle distribution in the bed agree well with experimental measurements. They demonstrate the practicality of the present hybrid method to solid particle sedimentation and bed formation behaviors of mixed as well as homogeneous particles.

Keywords: hybrid simulation method; multi-fluid model; discrete element method, sedimentation, bed formation

---

1. Introduction

In the latter stage of an unprotected loss-of-flow (ULOF) event, which is a representative initiator of a core disruptive accident (CDA) in sodium-cooled fast reactors (SFRs), molten core materials may relocate into a lower sodium plenum due to a gravity-driven discharge through potential paths such as control rod guide tubes. During this material relocation phase, the molten core materials are rapidly quenched and fragmented in the subcooled sodium and become solidified into smaller-sized particles [1]. The discharged particles, called debris, which are roughly 300 μm in Sauter mean diameter in SFR [2], settle down on a core support structure and/or an in-vessel retention (IVR) device such as a multi-layered debris tray installed in the bottom region of the reactor vessel and form debris beds [3].

In the safety analysis of CDAs, the sedimentation behavior of these particles is a crucial issue as it has a significant influence on heat removal from the fuel debris with decay heat. Numerous studies have focused on debris bed characteristics [4,5], but the shape of the debris bed is chosen arbitrarily, e.g., cylindrical [6], conical or Gaussian-shaped [7], heap-like [8,9], conical [10] or homothetically piled [11]. Moreover, the debris bed was assumed homogeneous and uniformly distributed above the lower-most level of the reactor pool as supposed in a preceding study of the cooling capability of a debris bed in a reactor accident [4].

Experimental investigations with simulant materials composed of homogeneous solid particles were conducted recently [12,13] and, in connection with this study, an exploration was performed in our previous studies for mixtures of particles with different properties [14,15]. These experimental analyses discovered the nature of the shape characteristics of the particle bed and its dependence on some important factors, such as particle density and diameter as well as the diameter and length of the nozzle, which is used to discharge solid particles into the water pool [12–15]. As the experimental process consumes a high cost and long duration, the evaluation of such sedimentation behaviors is performed knowing the limited scope of reproducibility. Although the high cost and long duration of such experiments can be reduced, it is difficult to imitate the reactor conditions rigorously. However, without experimental data, a justifiable extrapolation to reactor conditions would be tough. This is because numerical analysis, which reflects findings obtained by experiments with the addition of possible reproducibility of the reactor conditions, enables us some extrapolation, even though the experimental scope is limited. Consequently, we require the numerical study on sedimentation behavior using solid particles as simulated debris.

The first practical tool, the SIMMER-II code, was developed at the Los Alamos National Laboratory (LANL). This code was used in many experimental and reactor analyses [16,17]. Subsequently, SIMMER-III was introduced by the Japan Nuclear Cycle Development Institute (JNC), presently called the Japan Atomic Energy Agency (JAEA). The SIMMER-III code is a two-dimensional, multi-velocity-field, multiphase, multicomponent, Eulerian, fluid dynamics combined with a space- and energy-dependent neutron kinetics model [18]. The major drawback of SIMMER-III is its two-dimensional treatment, although the computational technology of SIMMER-III has improved the safety evaluation of liquid–metal-cooled fast reactors (LMFRs). The lack of dimensionality in SIMMER-III requires certain conservatism because of large uncertainties in accident analysis. Consideration of the three-dimensional distribution of core materials including neutron absorbers would reasonably simulate the reactivity effect, which could lead to severe re-criticality, caused by the relocation of disrupted core materials [19,20]. Although we need large computational resources in large-scale three-dimensional simulations of safety analyses, the parallelization technology will reduce the computational load.

Reasonable numerical simulations of the transient debris behavior involving the thermal-hydraulic phenomena of the surrounding fluid phases require a comprehensive computational tool, considering a complicated multiphase mechanism for the debris bed cooling capability. A reliable and trustworthy tool, the reactor safety analysis code, SIMMER-IV [19,20], has been applied to key phenomena such as fuel discharge and relocation [21] and pool sloshing [22], as well as safety analyses of the transition phase [19,20] and the post-disassembly expansion phase [23], in CDAs of an LMFR successfully. It is a three-dimensional multi-velocity-field, multiphase, multicomponent, Eulerian, fluid dynamics code coupled with a neutron kinetics model and a fuel pin model. This code has recently been productively applied to simulations of critical thermal-hydraulic phenomena in CDAs as well as to reactor safety analysis. However, mechanical interactions among solid particles and discrete phase characteristics of solid particles are not modeled directly in the associated fluid dynamics calculations in the simulations of multiphase flows with rich solid particles.

Appropriate calculations of the interaction between the fluid phases and particles, as well as the interactions between particles themselves, are necessary for simulating such phenomena reasonably. A discrete method can predict the solid particle interactions and motions directly based on the solid-phase continuity at a microscopic level. In principle, with proper initial and boundary conditions, Newton's equations of motions are solved together. Unlike conventional grid methods, it is not necessary to assume uniform constituency and constitutive relations for discrete solid particles [24]. Cundall and Strack [25] advanced and introduced the discrete element method (DEM), which is the most successful numerical approach among this category of numerical methods. Multi-body collisions can be calculated accurately with an explicit force model. Besides, DEM gives transient trajectories and velocities of particles as local information. Moreover, the fluid–particle interactions are calculated by coupling the DEM with computational fluid dynamics methods using drag force terms straightforwardly [26].

A theoretical model was established to explore the characteristics of the dense particle bed in simulations of the decay heat removal from the fuel debris formed in the lower plenum of SFRs [27]. The inter-particle interaction is anticipated to be a substantial phenomenon in the dense particle bed. Inter-particle collisions and contacts were demonstrated to be influential. In addition, this model is able to explain that characteristics of the debris movement initiated by sodium vapor flow in debris beds.

Guo et al. [28–31] combined DEM with the multi-fluid model and developed a hybrid computational method based on the theoretical background introduced above. It is expected that this numerical method reproduces the particle behavior involved in the thermal-hydraulic phenomena reasonably. A semi-implicit time factorization approach solves the governing equations of the fluid phases, whereas DEM calculates particle movements through the coupling algorithm. In their governing equations, the multiphases are then coupled via drag terms explicitly. The fundamental applicability of the developed hybrid method has been validated by simulating typical gas–solid fluidized beds [28], the self-leveling of particle beds in a rectangular pool [29] and gas–liquid particle three-phase flows [30] in two-dimensional systems. Recently, a three-dimensional numerical simulation was performed for the self-leveling of the particle beds in a cylindrical pool, and a fundamental validation of the developed hybrid method was demonstrated successfully [31]. In addition, a DEM-based numerical study on the sedimentation behavior of solid particles in two-dimensional systems was accomplished [32]. The primitive numerical simulation study was not coupled with the fluid flow and particle dynamics. Therefore, a numerical analysis in a three-dimensional system that covers both a mixture of particles as well homogeneous ones is essential.

The objective of the present numerical study is to demonstrate the practicality of the hybrid method to particle sedimentation and bed formation behaviors of mixed particles as well as homogeneous ones. In the present study, three-dimensional simulations were performed and compared with results obtained in a series of particle sedimentation experiments in which gravity-driven solid particles [12,15] are discharged into a quiescent cylindrical water pool. Although the experiments using simulant materials did not reproduce the particle sedimentation and bed formation behaviors that are expected to occur under reactor accident conditions, the hybrid method will be validated for fundamental characteristics of the behaviors, which were measured in the experiments performed under controlled conditions.

2. Mathematical Treatment

In the present study, we performed three-dimensional calculations for solid particle sedimentation and bed formation behaviors of mixed as well as homogeneous particles using the hybrid computational method, which combines DEM with the multi-fluid model of the SIMMER-IV code. A detained explanation of the method is given by Guo et al. [28–31].

In the multi-fluid model of the SIMMER-IV code, the governing equations of multi-fluid phases are the conservation equations of mass, momentum and energy, in terms of the local mean variables over a computational cell, in abbreviated form [19]:

$$\frac{\partial \overline{\rho}_m}{\partial t} + \nabla\cdot\left(\overline{\rho}_m \vec{v}_q\right) = -\Gamma_m,$$

(1)

$$\frac{\partial \overline{\rho}_q \vec{v}_q}{\partial t} + \sum_{m\in q} \nabla\cdot\left(\overline{\rho}_m \vec{v}_q \vec{v}_q\right) + \alpha_q \nabla p - \overline{\rho}_q \vec{g} + K_{qS}\vec{v}_q - \sum_{q'} K_{qq'}\left(\vec{v}_{q'} - \vec{v}_q\right) - \vec{VM}_q$$
$$= -\sum_{q'} \Gamma_{qq'}\left[H\left(\Gamma_{qq'}\right)\vec{v}_q - H\left(-\Gamma_{qq'}\right)\vec{v}_{q'}\right],$$

(2)

$$\frac{\partial \overline{\rho}_M e_M}{\partial t} + \sum_{m\in M} \nabla\cdot\left(\overline{\rho}_m e_m \vec{v}_q\right) + p\left[\frac{\partial \alpha_M}{\partial t} + \nabla\cdot\left(\alpha_M \vec{v}_q\right)\right] - \frac{\overline{\rho}_M}{\overline{\rho}_q}\left[\sum_{q'} K_{qq'}\left(\vec{v}_q - \vec{v}_{q'}\right)\left(\vec{v}_q - \vec{v}_{q'}\right)\right]$$
$$+ K_{qS}\vec{v}_q\cdot\left(\vec{v}_q - \vec{v}_{qS}\right) - \vec{VM}_q\cdot\left(\vec{v}_q - \vec{v}_{GL}\right) = Q_{N,M} + Q_{M,M} + Q_{H,M},$$

(3)

where subscripts m, q and M denote the density, velocity and energy component, respectively; t is the time; $\overline{\rho}_m$ and Γ_m are the macroscopic density and total mass transfer rate per unit volume from the density component m, respectively; $\vec{v}_q$ is the vector of the velocity field q; p denotes the pressure, $\vec{g}$ the gravitational acceleration, K_{qS} the momentum exchange function between the velocity field q and structure component, $K_{qq'}$ the momentum exchange function between the velocity field q and q', $\vec{VM}_q$ the virtual mass term of the gas phase, $H(x)$ the Heaviside unit function, $\Gamma_{qq'}$ the mass transfer rate from component q and q', and e_M and α_M are the specific internal energy and volume fractions of component M, respectively; Q_N, Q_M and Q_H are, respectively, the nuclear heating rate, the rate of energy interchange due to the mass transfer and other heat transfer rates. The subscript GL indicates terms such as the averaged velocity relevant at interfaces between the vapor and liquid.

In the DEM calculations, the solid phase is represented by a discrete phase. Newton's law [24] is used to describe the translational motion of a particle as

$$m_i \frac{d^2 \vec{r}_i}{dt} = \sum_j \vec{F}_{ij}^c + \vec{F}_i^f + \vec{F}_i^g,$$

(4)

$$\vec{v}_i = \frac{d\vec{r}_i}{dt},$$

(5)

$$\vec{I}_i \frac{d^2 \vec{\theta}_i}{dt^2} = \sum_j \vec{A}_{ij},$$

(6)

$$\vec{\omega}_i = \frac{d\vec{\theta}_i}{dt},$$

(7)

where m_i is the mass of particle i; $\vec{r}_i$ and $\vec{v}_i$ are the vectors of its position and velocity, respectively; $\vec{\theta}_i$ and $\vec{\omega}_i$ are the vectors of its angular displacement and angular velocity, respectively; $\vec{F}_{ij}^c$ is the contact force of particle i with neighboring particle j or wall; $\vec{F}_i^f$ the total solid–fluid interaction force on particle i; $\vec{F}_i^g$ is the gravitational force; $\vec{I}_i$ and $\vec{A}_{ij}$ and $\vec{\omega}_i$ are the moment of inertia, torque and angular velocity, respectively.

In the present DEM, which assumes that particles are spheres, a viscoelastic contact model [33] is applied to calculate the contact forces between the particles, as well as between particles and the wall. The DEM calculation is coupled with the fluid dynamics one of the SIMMER-IV code, which is based on a time factorization time-splitting approach, through the terms of solid–fluid interactions in the momentum conservation Equation (2). Time-step sizes for the fluid dynamics and DEM calculations are controlled independently. In the SIMMER-IV code, the Courant condition, the optimum pressure iteration condition and the excessive vaporization/condensation iteration condition are used mainly to optimize the time-step size of fluid dynamics calculations. For the DEM calculation, in which velocities and positions of particles are updated based on an explicit scheme, a large time-step size is limited so as to prevent unphysical disturbances of the particle in the particle–fluid interactions. In addition, the smallest time period, by which two particles reach their maximum overlap from the initial contact in the collision, is considered. In general, the latter time-step size is much smaller in the DEM calculations.

3. Verification Experiment and Numerical Simulations

3.1. Particle Sedimentation Experiments

A series of particle sedimentation experiments was performed for gravity-driven discharges of solid particles into a quiescent cylindrical water pool [12,15]. For the experimental setup (Figure 1), a transparent cylindrical tank filled with water with an inner diameter D_c of 375 mm and a height of 1040 mm is the main apparatus. Solid particles, which are kept in an overhead hopper initially, were poured into the water tank though a nozzle having an inner diameter d_n of 40 mm. The initial water level is 480 mm, and the exit of the nozzle is located at a height N_h of 473 mm from the bottom of the tank. In the experiments, the solid particles discharged from the nozzle fall into the water pool, and are deposited on the bottom of the tank finally forming a particle bed with a conical shape. The particle materials include Al_2O_3 and SS. For homogeneous spherical particles, their bulk volume is 5 L, whereas the two types of spherical particles with a bulk volume of 2.5 L are mixed up as a binary mixture of particles with different properties.

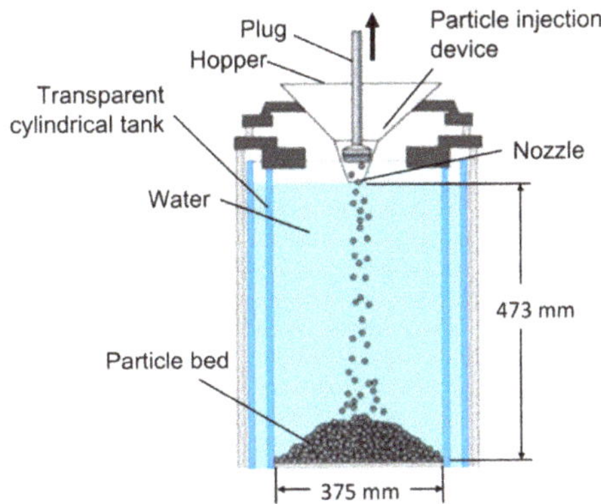

Figure 1. Schematic illustration of the setup for the particle sedimentation experiments.

3.2. Simulation Conditions

We calculated several cases of particle sedimentation experiments using binary mixtures of spherical Al_2O_3 and SS particles with different densities or different sizes as well as homogeneous particles. In these arrangements, we considered two cases with homogeneous particles and five cases with binary mixed particles. In the binary mixture, we chose three cases with different densities and sizes but set an equal volume mixing ratio. In addition, we took the setup as two cases with binary mixed particles of different volume mixing ratios. However, in this simulation process, we selected particles of larger sizes to avoid prolonged calculations times. The simulation conditions and particle properties of the selected cases are listed in Table 1. Two cases, labeled Cases 1 and 2, correspond to homogeneous solid particles and five types of binary mixtures, labeled Cases M1–M5, were chosen for the simulations presented. Figure 2 shows the initial position of the solid particles and fluid phases in the computational domain of Case M1. The number of computational cells for fluid phases is 25 × 25 × 25 in a X-Y-Z geometry, and the width of each cell is 24 mm. For solid particles, which filled the hopper initially, 31,568 DEM particles were used in simulations in the same bulk volume as the experiment.

Table 1. Simulation conditions and particle properties.

Homogeneous Particles (Bulk Volume Is 5.0 L)	
	Particle Type
Case 1	SS (ρ_p = 7800 kg/m^3; d_p = 6 mm)
Case 2	Al$_2$O$_3$ (ρ_p = 3600 kg/m^3; d_p = 6 mm)
Binary Mixture of Particles (Equal Volume Mixing Ratio)	
Particle type	SS (ρ_p = 8050 kg/m^3), Al$_2$O$_3$ (ρ_p = 3720 kg/m^3)
Case M1	Al$_2$O$_3$ and SS particles with d_p = 6 mm
Case M2	Al$_2$O$_3$ particles with d_p = 4 and 6 mm
Case M3	SS particles with d_p = 4 and 6 mm
Binary Mixture of Particles (Different Volume Mixing Ratio)	
Case M4	Al$_2$O$_3$ and SS particles with d_p = 6 mm with volume mixing ratio of 1:3
Case M5	Al$_2$O$_3$ and SS particles with d_p = 6 mm with volume mixing ratio of 3:1

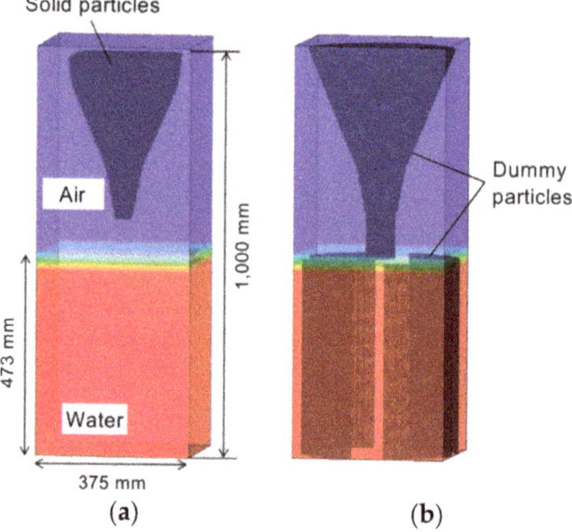

Figure 2. Schematic illustration of the setups for the simulation (color bar: volume fraction of water phase): (**a**) initial arrangement of solid particles; (**b**) arrangement of dummy particles.

Table 2 lists the model parameter values used in the DEM calculation. The time-step size was very small, as explained in Chapter 2, if real values (GPa scale) of Young's modulus E and the modulus of rigidity G are used in the calculation. To solve this problem, we refer the reader to [34]. Smaller artificial values (MPa scale or smaller) are applied to maintain the time-step sizes of order 10^{-5} s with sufficient accuracy. The particle hopper, which initially retains the solid particles, and the cylindrical tank are shaped by dummy DEM particles fixed in space to consider interactions with moving solid particles. The simulated wall surfaces of the hopper and the tank are not smooth and do not represent the real ones exactly. Although particle motions are affected in varying degrees by these artificially rough surfaces, especially of the inner wall of the nozzle, it is expected that the particle sedimentation and bed formation behaviors are not affected largely by the rough surface of the nozzle because the bulk particle flow in the nozzle is dense during the particle injection.

Table 2. Values of model parameters used in the DEM calculation.

Model Parameters	SS Particles	Al$_2$O$_3$ Particles	Dummy Particles
Poisson's ratio v [-]	0.305 (Cases 1, M1, M2, M4, M5) 0.21 (Case M3)	0.21	0.305 (Cases 1, M1, M2, M4, M5) 0.21 (Case M3)
Young's modulus E [Pa]	5.4×10^6 (Cases 1, M1) 5.0×10^5 (Case M3) 5.4×10^5 Cases M4, M5	9.0×10^6 (Cases 2, M1) 5.0×10^5 Case M2 9.0×10^5 Cases M4, M5	5.0×10^5
Modulus of rigidity G [Pa]	1.8×10^6 Cases 1, M1 5.0×10^5 (Case M3) 1.8×10^5 (Cases M4, M5)	3.0×10^6 Cases 2, M1 5.0×10^5 (Case M2) 3.0×10^5 (Cases M4, M5)	5.0×10^5
Dynamic friction coefficient ς [-]	0.25 (Cases 1, M1, M4, M5) 0.30 (Case M3)	0.25 (Cases 2, M1, M4, M5) 0.30 (Case M2)	0.20
Viscous damping coefficient γ_{nor} (normal direction) [s^{-1}]	1100 (Cases 1, M1, M4, M5) 7000 (Case M3)	1100	14,000
Viscous damping coefficient γ_{tan} (tangential direction) [s^{-1}]	1100 (Cases 1, M1, M4, M5)	1100	14,000

4. Results and Comparisons

4.1. Homogeneous Particles

Figure 3 shows visual snapshots of homogeneous particles falling in the water pool and their sedimentation behaviors at 1.0, 9.0 and 17 s after initiating particle injection for Case 1 and similarly at 1.0, 8.0 and 17 s for Case 2. In Case 1, the SS particles form a jet-like particle flow in the pool. From the figures, the lateral dispersion of the SS particles is most likely to be underestimated in the present simulations. This may be caused by the lack of a turbulence model in the present simulations. Nevertheless, although a wide-spreading behavior is observed in Case 2 for the lighter Al$_2$O$_3$ particles during their fall, the corresponding lateral dispersion of the Al$_2$O$_3$ particles in the water pool is underestimated in the simulation. Moreover, for Case 2, the turbulence flow induced by the falling particles may influence the particle motion and enhance the dispersal of the lighter Al$_2$O$_3$ particles, however the multi-fluid model used in the present simulation does not consider turbulence effects in the water, which will have an impact on the motion of the solid particles. In addition, in another experiment using Al$_2$O$_3$ particles with d_p = 6 mm in air, less lateral spreading of the particles is observed because turbulence effects on the solid particles were relatively small. Therefore, the difference in observations of the particle falling behavior may be caused by the lack of a turbulence model in the present simulation. Nevertheless, for both Cases 1 and 2, the shape of the bed formed after the completion of particle sedimentation is comparable in both simulations and experiments. The radial variation of the bed height of the homogeneous particles with d_p = 6 mm is presented in Figure 4; specifically, the experiment results, which did not show strong axial asymmetry in bed shapes, are compared with (a) the calculated radial bed height of SS particles, Case 1, and (b) the calculated radial bed height of Al$_2$O$_3$ particles, Case 2. In this figure, the side-view height was measured from the side-view shot image of the particle bed. A strong agreement is seen between the two sets of results for the side-view profile of both SS and Al$_2$O$_3$ particles. This is because even if the lateral spreading of the particles is smaller, particles settling higher up on the bed will tumble down the mound, and hence the bed will finally have a similar repose angle.

1.0 s   9.0 s   17 s   1.0 s   8.0 s   17 s

(**a**)   (**b**)

Figure 3. Falling and sedimentation behavior of homogeneous particles with $d_p = 6$ mm: (**a**) Case 1; (**b**) Case 2.

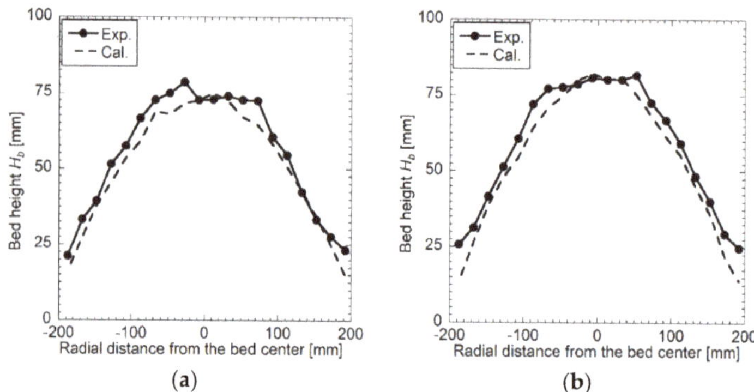

(**a**)   (**b**)

Figure 4. Side-view height of the homogeneous particle bed in the radial direction: (**a**) Case 1; (**b**) Case 2.

4.2. Mixed Particles

Visual snapshots of the falling and sedimentation behaviors of mixed particles in the water pool are similarly presented (Figure 5). A comparison of snapshots between simulation and experimental results is given at three instants in time after particle injection has begun for Cases M1–M5. In Case M1, the bulk particle jet flow is formed mainly by heavy SS particles, and lighter Al_2O_3 particles are dispersed widely during their fall in the pool. Similar falling behavior is observed in Case M2, for which larger falling Al_2O_3 particles with $d_p = 6$ mm spread in the pool, whereas smaller ones form a particle jet flow. In Case M3, particles dispersion from the main jet flow is not observed. However, for the particle mixtures with different volume mixing ratios, Cases M4 and M5, the dispersion of the lighter Al_2O_3 particles still occurs. Although in simulations the lateral dispersion of particles is likely underestimated during their fall compared with observations, the observed falling and sedimentation behaviors are reasonably well reproduced in these five cases.

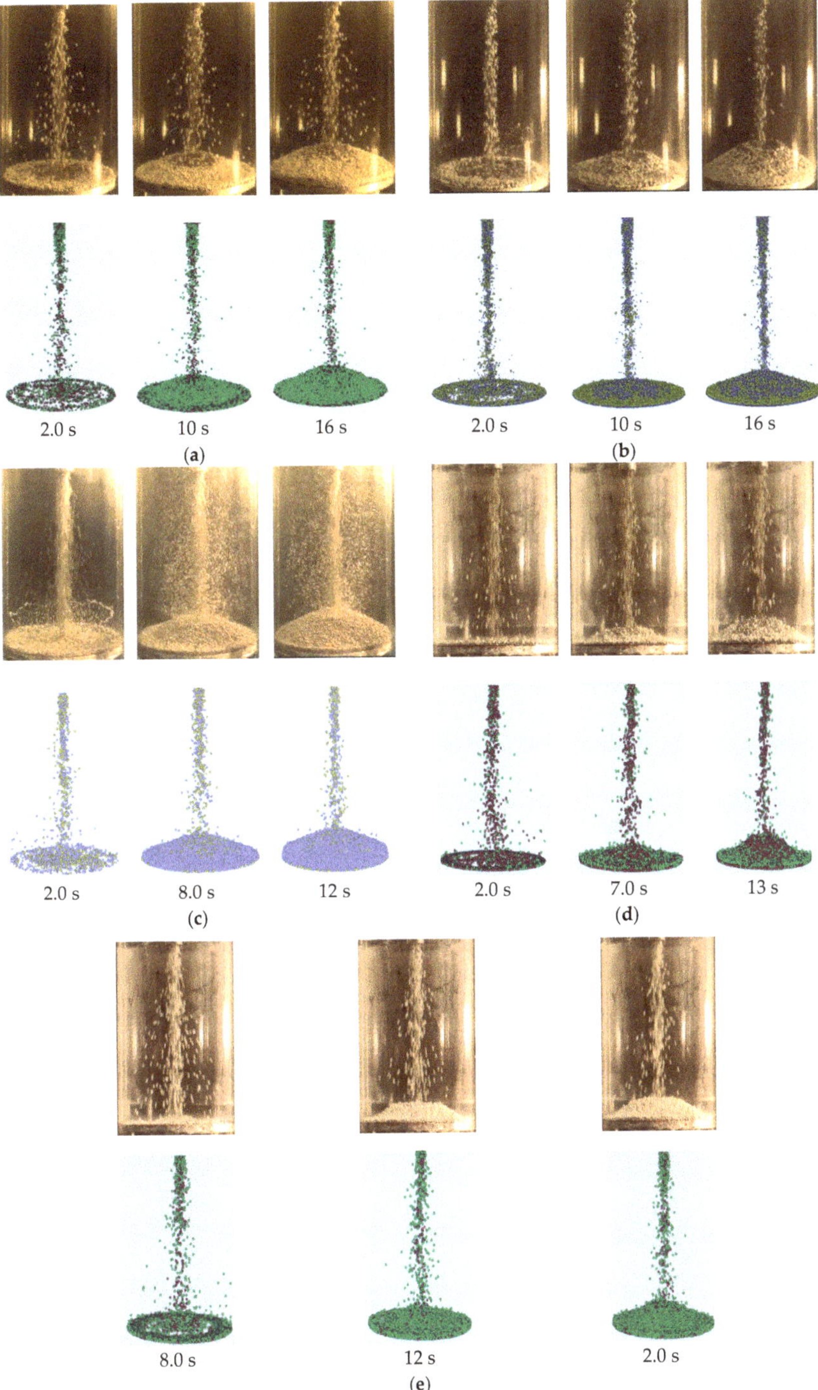

Figure 5. Falling and sedimentation behaviors of mixed particles with different properties: (**a**) Case M1; (**b**) Case M2; (**c**) Case M3; (**d**) Case M4; (**e**) Case M5.

In Figure 6, comparisons of the radial bed height variation after the completion of particle sedimentation are shown for the five mixed particle cases, in which strong axial asymmetry was not observed in the bed shapes. Although a general agreement between the simulation and experimental results is obtained, the slight discrepancies in bed height may be due to differences in falling and sedimentation behavior.

4.3. Analysis of Particles Distributions

In Cases M1 and M2, M4 and M5, after the particle sedimentation was completed, top-view pictures of particle beds were taken at four cross-sectional positions by removing particles from the bed using a vacuum suction device. In Figure 7, photographs are shown of the cross-sections at four axial positions for binary mixtures of Al_2O_3 and SS particles with different particle sizes in equal and different volume proportions. Here, the mixtures are (a) Al_2O_3 and SS particles with d_p = 6 mm in equal volume proportions, (b) Al_2O_3 particles with d_p = 4 and 6 mm in equal volume proportions, (c) Al_2O_3 and SS particles with d_p = 6 mm in a volume proportion of 1:3 and (d) Al_2O_3 and SS particles with d_p = 6 mm in a volume proportion of 3:1. The axial positions indexed as 1, 2, 3 and 4 are the top, middle and bottom of the conical bed mound, and the middle of the cylindrical bed basement, respectively. For Case M1 (Figure 7a), the heavier SS particles concentrate around the bed center at the axial four positions, whereas the light Al_2O_3 particles disperse largely during their fall and after their impact on the bed mound. This behavior was also observed in Cases M4 and M5 with mixtures of Al_2O_3 and SS particles (Figure 7c,d). A similar particles distribution behavior occurs in Case M2 (Figure 7b), where larger or heavier Al_2O_3 particles are seen more in the periphery of the bed, although the behaviors of the dispersion and distribution do not substantially contrast those of Cases M1, M4 and M5. Nevertheless, the observed particles distribution behaviors were reproduced reasonably well by the present simulation in all cases.

The apparent areas of Al_2O_3 and SS particles mixed with different sizes and densities on the cross-sections at the four axial positions are shown in Figure 8, specifically (a) Case M1: Al_2O_3 and SS particles with d_p = 6 mm, (b) Case M2: Al_2O_3 particles with d_p = 4 and 6 mm, (c) Case M4: volume mixing ratio 1:3, Al_2O_3 and SS particles with d_p = 6 mm and (d) Case M5: volume mixing ratio 3:1, Al_2O_3 and SS particles with d_p = 6 mm. The axial positions of the cross-sections are the same as those in Figure 7. In this figure, the proportions of Al_2O_3 and SS particles were defined as the ratios of apparent areas of Al_2O_3 and SS particles to the cross-sectional area of the bed. In the experiments, the apparent areas of the particles in different colors were measured from the cross-sectional shot images of the bed using an image analysis tool. The ratios of the two different particles obtained in the simulations and the experiments are indicated in the upper and lower sides, respectively, at each axial position. A strong agreement between the simulated results of the particles distributed area at different axial positions with the experimental observations was established (Figure 8a–d). In Case M1, the central and lower parts of the bed are mainly occupied by SS particles and this behavior is seen also in Case M4. However, in Case M2, the ratio of the smaller particles in the bed increases in the height direction. Moreover, the lighter Al_2O_3 particles are seen in all parts of the bed in Case M5. Overall, the particles distribution behavior of the lighter and larger Al_2O_3 particles observed in Cases M1, M2, M4 and M5 is reproduced quantitatively in the present simulations.

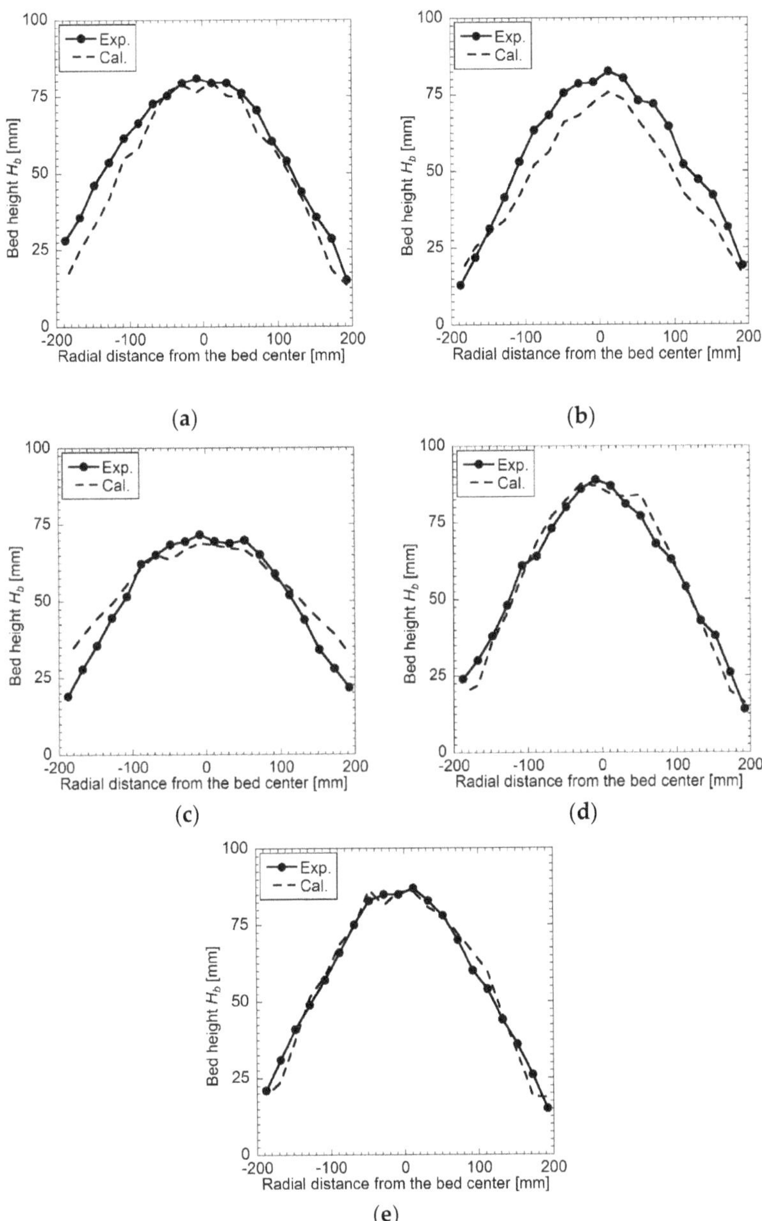

Figure 6. Radial variation of bed height for mixed particles with different properties in the radial direction: (**a**) Case M1; (**b**) Case M2; (**c**) Case M3; (**d**) Case M4; (**e**) Case M5.

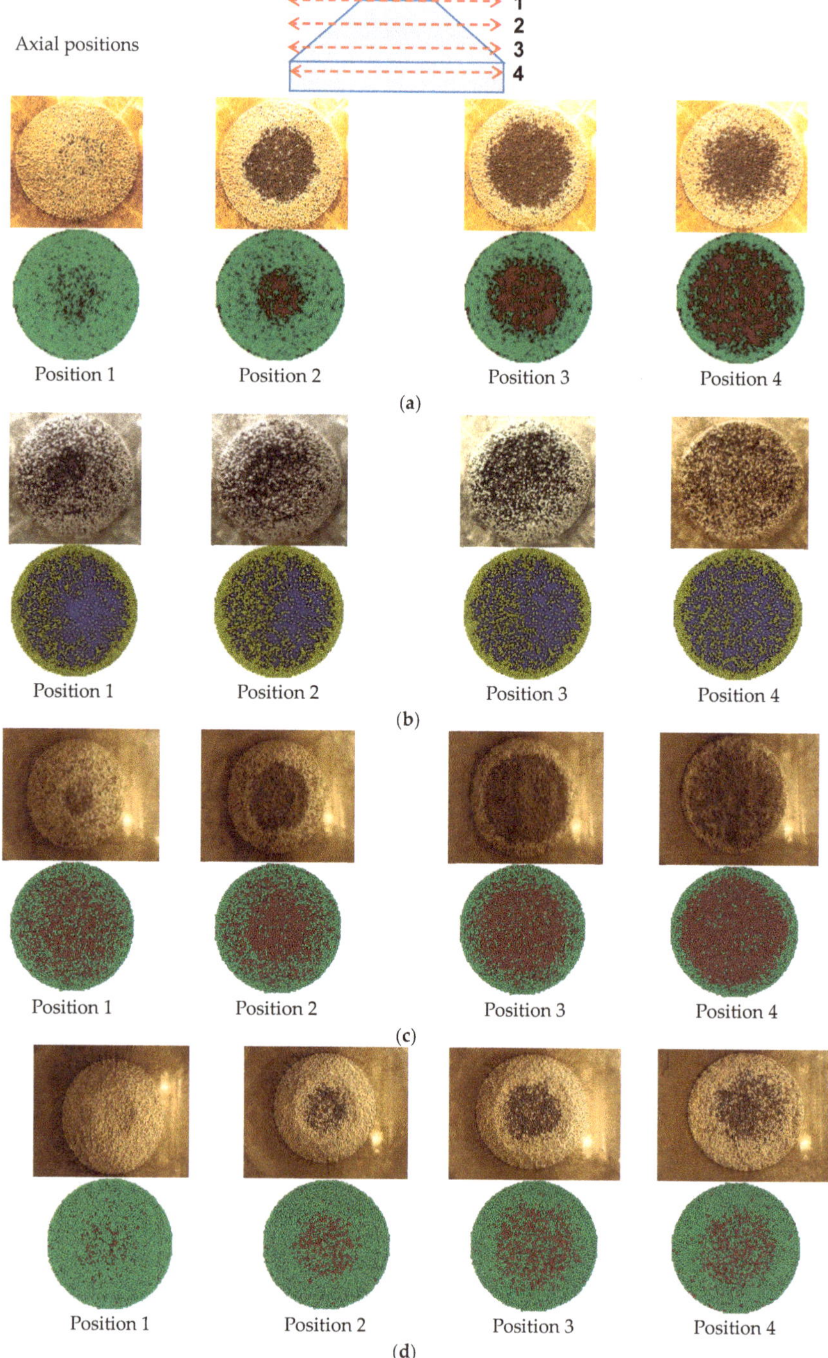

Figure 7. Top-view of mixed particle bed at four cross-sectional positions: (**a**) Case M1; (**b**) Case M2; (**c**) Case M4; (**d**) Case M5.

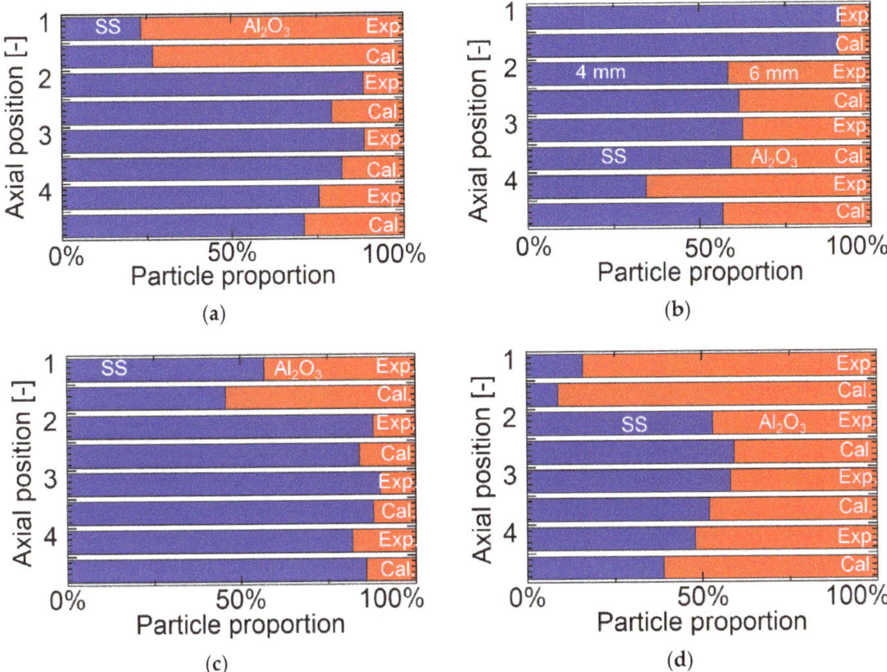

Figure 8. Particle distribution at four axial positions in the mixed particle bed: (**a**) Case M1; (**b**) Case M2; (**c**) Case M4; (**d**) Case M5.

5. Conclusions

In the present validation study of the hybrid numerical simulation method, three-dimensional simulations were performed and compared with results obtained in a series of particle sedimentation experiments of gravity-driven discharges of solid particles, the simulants of core debris, into a quiescent water pool. The governing equations of the multi-fluid phases were calculated using the conventional multi-fluid model, whereas for solid particles, the equations for the translational motion were calculated considering contacts among particles based on DEM. The simulation results on the particle falling behavior, bed height and particle distribution in the bed agree well with the experimental results for binary mixtures of particles with different densities or sizes as well as homogeneous particles. The particles distribution behavior of the lighter or larger particles in the binary mixtures was successfully reproduced using the present hybrid method. Although the lack of a turbulence model may underestimate the dispersion behavior of lighter and larger Al_2O_3 particles observed during their fall in the pool, the results of the present simulation demonstrate the practicality of the present hybrid method to the solid particle sedimentation and bed formation behaviors of both mixed and homogeneous particles. It might be of interest to consider quantifying the effect of the rough surfaces of the hopper and the tank walls, which were modeled by rather large dummy DEM particles, on the motion of solid particles in future work. Further validation is needed of the present method for a wide range of experimental conditions such as particle size and density, that have a strong influence on particle sedimentation and bed formation behaviors.

Author Contributions: Conceptualization, K.M., T.S. and K.K.; methodology, K.M., T.S. and K.K.; software, X.L. and L.G.; validation, T.M.; formal analysis, M.A.R.S.; investigation, M.A.R.S. and T.M.; resources, K.M. and T.M.; data curation, M.A.R.S.; writing—original draft preparation, M.A.R.S.; writing—review and editing, K.M.; visualization, M.A.R.S.; supervision, K.M. and T.M.; project administration, K.M., T.S. and K.K.; funding acquisition, K.M., T.S. and K.K. All authors have read and agreed to the published version of the manuscript.

Funding: This research was funded by the Japan Society for the Promotion of Science (JSPS) KAKENHI [Grant No. 16K06960] and Japan Atomic Energy Agency.

Acknowledgments: The authors would like to thank R. Kawata and Y. Ohara for assistance with the numerical simulations. We also thank E. Son for assistance with the experiments. Finally, we are grateful to Richard Haase, from Edanz Group (www.edanzediting.com/ac) for editing a draft of this manuscript.

Conflicts of Interest: The authors declare no conflict of interest. The funders had no role in the design of the study; in the collection, analyses, or interpretation of data; in the writing of the manuscript, or in the decision to publish the results.

References

1. Magallon, D.; Hohmann, H.; Schins, H. Pouring of 100-kg-Scale Molten UO_2 into sodium. *Nucl. Technol.* **1991**, *98*, 79–90. [CrossRef]
2. Fauske, H.K.; Koyama, K. Assessment of Fuel Coolant Interactions (FCIs) in the FBR Core Disruptive Accident (CDA). *J. Nucl. Sci. Technol.* **2002**, *39*, 608–614. [CrossRef]
3. Nakai, R.; Suzuki, T.; Kamiyama, K.; Seino, H.; Koyama, K.; Morita, K. Development of Level 2 PSA methodology for sodium-cooled fast reactors. (1) Overview of evaluation technology development. In Proceedings of the 8th International Topical Meeting on Nuclear Thermal Hydraulics, Operation and Safety (NUTHOS-8), Shanghai, China, 10–14 October 2010. N8P0095.
4. Yakush, S.; Kudinov, P.; Dinh, T.N. Modeling of two-phase natural convection flows in a water pool with a decay-heated debris bed. In Proceedings of the International Congress on Advances in Nuclear Power Plants (ICAPP 2008), Anaheim, CA, USA, 8–12 June 2008; pp. 1141–1150.
5. Ma, W.; Dinh, T.N.; Buck, M.; Burger, M. Analysis of the effect of bed inhomogeneity on debris coolability. In Proceedings of the 15th International Conference on Nuclear Engineering (ICONE15), Nagoya, Japan, 22–26 April 2007. ICONE15-10752.
6. Burger, M.; Buck, M.; Schmidt, W.; Walter, W. Validation and application of the WABE Code: Investigation of Constitutive Laws and 2D effects on Debris Coolability. *Nucl. Eng. Des.* **2006**, *236*, 2164–2188. [CrossRef]
7. Yakush, S.; Kudninov, P. Simulation of ex-vessel debris bed formation and coolability in a LWR severe accident. In Proceedings of the Implementation of Severe Accident Management Measures (ISAMM 2009), Schloss Böttstein, Switzerland, 26–28 October 2009.
8. Karbojian, A.; Ma, W.M.; Kudinov, P.; Dinh, T.N. A scoping study of debris bed formation in the DEFOR test facility. *Nucl. Eng. Des.* **2009**, *239*, 1653–1659. [CrossRef]
9. Weimin, M.A.; Truc-Nam, D. The effects of debris bed's prototypical characteristics on corium coolability in a LWR severe accident. *Nucl. Eng. Des.* **2010**, *240*, 598–608.
10. Zhang, B.; Harada, T.; Hirahara, D.; Matsumoto, T.; Morita, K.; Fukuda, K.; Yamano, H.; Suzuki, T.; Tobita, Y. Self-leveling onset criteria in debris bed. *J. Nucl. Sci. Technol.* **2010**, *47*, 384–395. [CrossRef]
11. Alvarez, D.; Amblard, M. Fuel levelling. In Proceedings of the 5th Information Exchange Mtg. on Post Accident Debris Cooling, Karlsruhe, Germany, 28–30 July 1982; pp. 9–12.
12. Shamsuzzaman, M.; Horie, T.; Fuke, F.; Kai, T.; Zhang, B.; Matsumoto, T.; Morita, K.; Tagami, H.; Suzuki, T.; Tobita, Y. Experimental evaluation of debris bed characteristics in particulate debris sedimentation behaviour. In Proceedings of the 21st International Conference on Nuclear Engineering (ICONE21), Chengdu, China, 29 July–2 August 2013.
13. Shamsuzzaman, M.; Horie, T.; Fuke, F.; Kai, T.; Zhang, B.; Matsumoto, T.; Morita, K.; Tagami, H.; Suzuki, T.; Tobita, Y. Experimental study on debris bed characteristics for the sedimentation behavior of solid particles used as simulant debris. *Ann. Nucl. Energy* **2018**, *111*, 474–486. [CrossRef]
14. Sheikh, M.A.R.; Son, E.; Kamiyama, M.; Morioka, T.; Matsumoto, T.; Morita, K.; Matsuba, K.; Kamiyama, K.; Suzuki, T. Experimental investigation on characteristics of mixed particle debris in sedimentation and bed formation behavior. In Proceedings of the 11th International Topical Mtg. on Nuclear Reactor Thermal Hydraulics, Operation and Safety (NUTHOS-11), Gyeongju, Korea, 9–13 October 2016.
15. Sheikh, M.A.R.; Son, E.; Kamiyama, M.; Morioka, T.; Matsumoto, T.; Morita, K.; Matsuba, K.; Kamiyama, K.; Suzuki, T. Sedimentation behavior of mixed solid particles. *J. Nucl. Sci. Technol.* **2018**, *55*, 623–633. [CrossRef]
16. Bohl, W.R.; Luck, L.B. *SIMMER-II: A Computer Program for LMFBR Disrupted Core Analysis*; LA-11415-MS; Los Alamos National Laboratory: Los Alamos, NM, USA, 1990.

17. Smith, L.L.; Bell, C.R.; Bohl, W.R.; Luck, L.B.; Wehner, T.R.; DeVault, G.P.; Parker, F.R. The SIMMER Program: Its Accomplishment. In Proceedings of the International Top. Mtg. on Fast Reactor Safety, Knoxville, TN, USA, 21–25 April 1985.

18. Tobita, Y.; Kondo, S.; Yamano, H.; Morita, K.; Maschek, W.; Coste, P.; Cadiou, T. The development of SIMMER-III, an advanced computer program for LMFR safety analysis and its application to sodium experiments. *Nucl. Technol.* **2006**, *153*, 245–255. [CrossRef]

19. Yamano, H.; Fujita, S.; Tobita, Y.; Sato, I.; Niwa, H. Development of a three-dimensional CDA analysis code: SIMMER-IV and its first application to reactor case. *Nucl. Eng. Des.* **2008**, *283*, 67–73. [CrossRef]

20. Yamano, H.; Tobita, Y.; Fujita, S.; Maschek, W. First 3-D calculation of core disruptive accident in a large-scale sodium-cooled fast reactor. *Ann. Nucl. Energy* **2009**, *36*, 337–343. [CrossRef]

21. Yamano, H.; Tobita, Y. Experimental analyses by SIMMER-III on duct-wall failure and fuel discharge/relocation behavior. *Mech. Eng. J.* **2014**, *1*, TEP0028. [CrossRef]

22. Yamano, H.; Suzuki, T.; Tobita, Y.; Matsumoto, T.; Morita, K. Validation of the SIMMER-IV severe accident computer code on three-dimensional sloshing behavior. In Proceedings of the 8th Japan-Korea Symposium on Nuclear Thermal Hydraulics and Safety (NTHAS-8), Beppu, Japan, 9–12 December 2012.

23. Onoda, Y.; Matsuba, K.; Tobita, Y.; Suzuki, T. Preliminary analysis of the post-disassembly expansion phase and structural response under unprotected loss of flow accident in prototype sodium cooled fast reactor. *Mech. Eng. J.* **2017**, *4*. [CrossRef]

24. Zhu, H.P.; Zhou, Z.Y.; Yang, R.Y.; Yu, A.B. Discrete particle simulation of particulate systems: Theoretical developments. *Chem. Eng. Sci.* **2007**, *62*, 3378–3396. [CrossRef]

25. Cundall, P.A.; Strack, O.D.L. A discrete numerical model for granular assemblies. *Geotechnique* **1979**, *29*, 47–65. [CrossRef]

26. Tsuji, Y.; Kawaguchi, T.; Tanaka, T. Discrete particle simulation of two dimensional fluidized bed. *Powder Technol.* **1993**, *77*, 79–87. [CrossRef]

27. Tagami, H.; Cheng, S.; Tobita, Y.; Morita, K. Model for particle behavior in debris bed. *Nucl. Eng. Des.* **2018**, *328*, 95–106. [CrossRef]

28. Guo, L.; Morita, K.; Tobita, Y. Numerical simulation of gas-solid fluidized beds by coupling a fluid-dynamics model with the discrete element method. *Ann. Nucl. Energy* **2014**, *72*, 31–38. [CrossRef]

29. Guo, L.; Morita, K.; Tagami, H.; Tobita, Y. Numerical simulation of a self-leveling experiment using a hybrid method. *Mech. Eng. J.* **2014**, *1*. [CrossRef]

30. Guo, L.; Morita, K.; Tobita, Y. Numerical simulations of gas-liquid-particle three-phase flows using a hybrid method. *J. Nucl. Sci. Technol.* **2016**, *53*, 271–280. [CrossRef]

31. Guo, L.; Morita, K.; Tobita, Y. Numerical simulations on self-leveling behaviors with cylindrical debris bed. *Nucl. Eng. Des.* **2017**, *315*, 61–68. [CrossRef]

32. Shamsuzzaman, M.; Zhang, B.; Horie, T.; Fuke, F.; Matsumoto, T.; Morita, K.; Tagami, H.; Suzuki, T.; Tobita, Y. Numerical simulation on sedimentation of solid particles used as simulant fuel debris. *J. Nucl. Sci. Technol.* **2014**, *51*, 681–699. [CrossRef]

33. Balevičiusa, R.; Kačianauskas, R.; Mróz, Z.; Sielamowicz, I. Discrete particle investigation of friction effect in filling and unsteady/steady discharge in three-dimensional wedge-shaped hopper. *Powder Technol.* **2008**, *187*, 159–174. [CrossRef]

34. Dziugys, A.; Peters, B. An approach to simulate the motion of spherical and non-spherical fuel particles in combustion chambers. *Granul. Matter.* **2001**, *3*, 231–265. [CrossRef]

Article

CFD and PIV Investigation of a Liquid Flow Maldistribution across a Tube Bundle in the Shell-and-Tube Heat Exchanger with Segmental Baffles

Grzegorz Ligus [1,*], Marek Wasilewski [2], Szymon Kołodziej [1] and Daniel Zając [3]

[1] Faculty of Mechanical Engineering, Opole University of Technology, 45-758 Opole, Poland; s.kolodziej@po.edu.pl

[2] Faculty of Production Engineering and Logistics, Opole University of Technology, 45-758 Opole, Poland; m.wasilewski@po.opole.pl

[3] Engineering and R&D Department, Kelvion Sp. z o. o., 45-641 Opole, Poland; daniel.zajac@kelvion.com

* Correspondence: g.ligus@po.edu.pl

Received: 31 July 2020; Accepted: 28 September 2020; Published: 2 October 2020

Abstract: The paper presents the results of research on liquid flow maldistribution in the shell side of a shell-and-tube heat exchanger (STHE). This phenomenon constitutes the reason for the formation of the velocity reduction area and adversely affects heat transfer and pressure drop. In order to provide details of the liquid distribution in STHE, two visualization methods were utilized. First, computational fluid dynamics (CFD) code coupled with the k-ε model and the laser-based particle image velocimetry (PIV) technique was applied. The tests were carried out for a bundle comprising 37 tubes in an in-line layout with a pitch $d_z/t = 1.5$, placed in a shell with $D_{in} = 0.1$ m. The STHE liquid feed rates corresponded to Reynolds numbers Re_{in} equal to 16,662, 24,993, and 33,324. The analysis demonstrated that the flow maldistribution in the investigated geometry originates the result of three main streams in the cross-section of the shell side: central stream, oblique stream, and bypass stream. For central and oblique streams, the largest velocity reduction areas were formed in the wake of the tubes. On the basis of the flow visualization, it was also shown that the in-line layout of the tube bundle helps to boost the wake region between successive tubes in a row. Additionally, unfavorable vortex phenomena between the last row of tubes and the lower part of the exchanger shell were identified in the investigations. The conducted studies confirmed the feasibility of both methods in the identification and assessment of fluid flow irregularities in STHE. The maximum error of the CFD method in comparison to the experimental methods did not exceed 7% in terms of the pressure drops and 11% in the range of the maximum velocities.

Keywords: CFD; PIV; shell-and-tube; shell side; tube bundle; heat exchanger; baffle; maldistribution

1. Introduction

Shell-and-tube heat exchangers (STHEs) find widespread applications in industry. They are utilized not only in areas whose purpose is only heat exchange but also in devices that are applied to perform other processes in technology. Consequently, the performance of many processes is dependent on the operating parameters of heat exchangers. The improvement of the operating parameters of heat exchangers is associated mainly with the need to perform geometric modifications sometimes leading to the comprehensive redesign of heat exchangers and thus the development of new types of devices. Another direction of modifications can be associated with material use that ensures better heat transfer parameters coupled with adequate strength parameters of heat exchangers. Extensive research is also carried out in the areas concerned with selecting optimal parameters of fluid flow in the heat exchanger.

One of the first significant modifications to the STHE was associated with the use of segmental baffles in the shell side. The use of baffles leads to the fluid flow in such a way that its turbulence is increased, which directly contributes to the improvement of the performance of the heat exchangers. Although new heat exchanger designs have emerged in recent years, STHEs are still a standard solution in many cases and in some cases forms the only viable solution. Today, the development research of industrial flow equipment is most often carried out in two stages. Initially, analysis using computational fluid dynamics (CFD) is performed, whose main purpose involves the selection the most suitable alternative of the investigated parameters among many possibilities. Then, experimental verification of selected parameters is performed. In this area, in particular, methods based on optical techniques are recognized in all fields of engineering and contribute to a significant increase in the reliability of the results gained on the basis of CFD codes. One of the most popular techniques currently utilized for this purpose is particle image velocimetry (PIV). It is one of the laser-based methods that allows the representation and analysis of the velocity fields in an investigated region. As a consequence, PIV offers a complementary tool to the application of CFD. Numerical and experimental investigations of classical STHEs with baffles are often undertaken in a variety of papers (see the following sections). However, little research has dealt with identifying flow structures and examining flow maldistribution across the tube bundle. The aim of this paper is to demonstrate the role taken by the adequate selection of flow parameters in the generation of variations of the local flow velocity in the device, which in turn led to flow maldistribution. In addition, the results should serve to provide assistance to an extensive group of researchers who carry out CFD tests in flow systems containing tube bundles in a variety of devices.

1.1. Insights from Literature in the Field of CFD Application in Heat Exchangers

Following the rapid development of computer techniques, research applying CFD methods is becoming increasingly common. Due to a number of advantages (e.g., low research costs, easy design modifications in the investigated devices, access to the common results), this method can provide the ready solution to the ever increasing economic and technical demand for devices applied in chemical and process engineering. The CFD method has been successfully applied on many occasions in research concerned with the description of the flow and heat transfer phenomena and the optimization of heat exchangers. When we perform an analysis of research on using the CFD method in heat exchangers, we can make particular reference to the study reported by Bhutta et al. [1], which forms an overview of the studies carried out by application of this method. The authors state that the CFD method was applied in the studies involving fluid flow maldistribution, pressure drop and thermal analysis, fouling, and in the design and optimization phases. A variety of interesting results can be found in the area of plate-fin heat exchangers, where flow maldistribution forms one of the major factors affecting the decrease of the effective heat exchange [2–4].

Kim et al. [5] performed an assessment of the types of flow patterns in STHE. For this purpose, three types of headers were investigated: A, B, and C. The conclusion is that the design with the longest fin offers the best performance. Ozden et al. [6] carried out a study by application of the same type of the heat exchanger. The conclusion section contains a statement that few recirculation regions are formed at the rear of the baffles when the number of baffles is small.

A large proportion of research works takes on the subjects concerned with the aspects of pressure drop reduction. Wang and Dong [7] conducted the research on various types of supporting structures applied for STHEs. The authors concluded that their geometry forms an important aspect in terms of the performance of the devices. As a result of the comparison of six types of supporting structures, it was demonstrated that rod elements are the most reliable and suitable supporting structure with a minimum pressure drop.

The study by Wang et al. [8] contains a statement that the pressure drop of a combined multiple shell-pass shell-and-tube heat exchanger tends to be lower than shell-and-tube heat exchanger with segmental baffles, for the same heat transfer conditions by almost 13%. Mohammadi et al. [9] demonstrated that the heat exchanger comprising a horizontal baffle has a 20% higher heat transfer coefficient combined with 250% greater pressure drop than the one with vertical baffles. Other examples of works concerned with STHEs and focusing on heat exchange aspects include articles Ozden and Tari [6] and Raj and Velraj [10].

A noteworthy research is reported in Sun et al. [11]. The authors investigated the effects of the application of inclined trefoil-hole baffles on the performance of STHEs. For this purpose, a comparison of this design with a similar model of the heat exchanger equipped with segmented baffles was performed. The comparison of numerical calculations demonstrated that the heat transfer coefficient in the exchanger with inclined trefoil-hole baffles is lower by 23.89% than in the exchanger with standard segmented baffles. However, a significant decrease in pressure drop was observed (pressure drop decrease was equal to 44.19%), which resulted in the higher specific value of the heat transfer coefficient related to the pressure drop by 36.38% than in the traditional heat exchanger design. The literature also contains studies on fouling in various types of heat exchangers (e.g., Jun and Puri [12], De Bonis and Ruocco [13]).

A detailed analysis of the studies available in the field of CFD application and concerned with heat exchangers indicates that several models can be applied to successfully represent turbulent flow. The obtained experimental results have repeatedly confirmed the compliance of CFD applications with the results of other research techniques in the validation processes. The standard k-ε model forms the most commonly model applied in the literature. The use of this model ensures high compliance of CFD test results with experimental tests (the range of the error in the analyzed works was from 2% to 10%) with the minimum possible load of computational units [1].

1.2. Insights from the Literature in the Field of PIV Application in Heat Exchangers

The common applications of the PIV method in the studies concerned with heat exchangers often involve studies concerned with the effect of the geometry of individual structural elements on either hydrodynamics or aerodynamics of the fluid flow. Such issues were raised e.g., by Iwaki et al. [14]. Using the PIV method, the study investigated the variations in the hydrodynamic characteristics of liquid flow across the tube bundle depending on the layout of the tubes in the bundle. The experiments were performed for in-line and staggered bundles with a pitch to diameter ratio of 1.5. The research demonstrated that parameters such as velocity distribution, flow structures, wake, and turbulent structures in the tube bundles should be taken into account when heat exchangers are designed, as formation of these structures is considerably dependent on the geometrical conditions in which fluid flow occurs. During the measurements, it was noted that turbulence for the staggered system behind the second row of pipes increases to a greater extent than for the in-line system. It was also confirmed that the flow in the staggered system was more homogeneous, since such geometry results in the mixing of flow streams and stabilization of the flow structure over a shorter distance. The extension of the above research to be applied with regard to the conditions of two-phase gas-liquid flow can be found in study: Vertical, bubbly, cross-flow characteristics over tube bundles [15]. In a study conducted in a system with a similar geometry, it was found that despite the fact that the difference in turbulence intensity between two configurations decreased in comparison to two-phase and single-phase conditions, the degree of turbulence in a staggered system still remains higher than in an in-line system. An important conclusion contained in this publication also includes the confirmation of the possibility of the application of computer image analysis in the measurement of velocities of both the liquid and gas phases and in the determination of the void fractions.

More complex shell-side geometry was previously investigated (e.g., Zhang et al. [16]), where the investigation involved the distribution of oil flow in a disc-type transformer winding. Despite the difficult imaging, it was noted that total volume of liquid derived from the PIV measurements is constantly slightly higher compared to the results gained by measurements using a flowmeter. The authors also pointed out that the end-wall effects in such geometry are usually neglected. We can note that such effects should be considered important in the experimental research concerned with various types of shell-side designs in heat exchangers.

The first attempt of visualizing PIV flow in an industrial heat exchanger was reported in paper Planar PIV experiments inside a transparent shell-and-tube exchanger [17]. This investigation involved the classic STHE with an internal diameter of 0.301 m and a bundle of 54 tubes with a diameter of 0.2 m. The authors carried out their research in their subsequent works (Turbulent flow in a no-tube-in-window shell-and-tube heat exchanger [18], Two-Phase Flow Regimes in Exchangers and Piping: Part 1. [19]), where a description was provided of the case of the shell side with four segmental baffles, for which the characteristic velocity contours were determined in different span-wise locations. It was indicated in the study that individual streamlines are superimposed as well as stagnation points are identified. In the research, it was clearly demonstrated that in the case of testing of flows in as complex geometries as in the case of shell side, it is very important to combine experimental research and CFD.

The literature also includes research on fluid distribution systems in heat exchangers. Tests of this type were carried out in the shell side as well as in the tube side. Wang et al. [20] investigated the effects of the introduction of two types of porous baffles installed with the purpose of regulating fluid distribution in the inlet part to the tube side and in the splitter bar to the tubes. It was found that the use of porous baffles can lead to a significant improvement of the fluid distribution. The comparison of the straight and arc baffle demonstrated that the latter offers better performance of the rectification process. In contrast, the introduction of a splitter bar into the tubes effectively reduces vortex formation in the tube side.

The issues associated with improvement of fluid distribution in heat exchanger tubes were also reported in article PIV measurement of flow structures in a circular heat exchange tube with central slant rod inserts [21]. The authors determined the liquid flow velocity fields in a circular heat exchange tube with central slant rod inserts using the stereoPIV method and additionally, on the basis of CFD simulations, the characteristics of heat transfer were determined. This study also demonstrated that the number of generated vortices is relative to the geometry of the central slant rod and the vortex intensity increases with the increase of the Reynolds number. The uncertainty of the velocity field measurement was also estimated, and it was demonstrated that the maximum relative errors for the velocities in the x, y and z directions are equal to approximately 3%, 6%, and 1.5%, respectively. The work also included a comparison of the results obtained by CFD simulations and experiment results using the stereoPIV technique, and it was found that the deviations of vorticity and z-velocity distribution are found within the range of 8% and 5%, respectively. Therefore, the study contains a conclusion that the overall efficiency of heat transfer in tubes equipped with central slant rod inserts is reliable.

The analysis of fluid flow in the shell side was also investigated by Lee at al. [22] and Delgado at al. [23]. The two works report the results of flow visualization using the PIV method in a helical coil steam generator. In the research, the authors observed several dark regions resulting from the difficulties of applying the PIV method for the analysis in the shell side. In order to improve the measurement methodology in the work of Delgado at al. [24], the optical system was coupled with a Scheimpflug module and a computer image correction was applied, and consequently, the image was converted from trapezoidal to rectangular. The authors noted, however, that the use of the Scheimpflug module resulted in a maldistribution of the light and led to shadow formation in the tube bundle. Hence, it was therefore necessary to use a Gaussian distribution for the color scale and the image brightness was increased by 20% for the purposes of obtaining better identification of seeding particles on this side of the image, where illumination was unsatisfactory. Besides, Im at al. [25] discussed obstacles faced during imaging process in the shell side. In the case of large field of view, a greater part of the area that could be potentially applied for flow analysis of the shell side was obstructed by horizontal heating rods. Consequently, the velocity vectors in these areas were found to be incorrect. This problem did not occur in the case of local measurements using small fields of view.

The effect of segmented baffles on the STHE performance was analyzed by Chang et al. in [26]. Tests were performed by application of a heat exchanger with two baffles located along a tube bundle. Velocity profiles between the baffles for co-current and countercurrent flow were determined for this purpose. CFD simulations were also performed, the results of which were identical to the reported PIV experiment. The paper also discussed the effect resulting from the presence of an intense liquid stream in the bottom part of the shell side. This phenomenon was linked to clearance between the baffle and the shell of heat exchanger. During the research, a zone of large circulation behind the baffle was also identified. There were also differences in the liquid distribution depending on the technique applied to feed fluid into the heat exchanger.

A testing procedure carried out along the tube bundle for the case of a heat exchanger comprising helicoidal baffles was reported by Wen et al. [27]. In this case, researchers conducted an experiment with purpose of comparing standard plain helical baffles with an improved fold helical baffle design in which a triangular gap forming between adjacent baffles was embedded. The upgraded baffle made it possible to obtain better heat transfer characteristics coupled with virtually unchanged pressure drops. The performance of such a heat exchanger increased in the same flow conditions in the range from 2% to 8%, and the results correlated very well with the analysis of velocity field changes obtained on the basis of PIV tests.

In the summary of the literature in the field of PIV research concerned with heat exchangers, we can note the widespread use of this technique. The analyzed cases apply a wide range of geometries and flow parameters and in the conditions combined with adequately selected measurement methodology, the results derived from PIV measurements are well simulated by means of computer techniques as well as other experimental measurement techniques.

1.3. The Research Gap and the Adopted Target of the Study

As we can see from Sections 1.1 and 1.2, studies of the flow maldistribution in STHEs have so far been carried out mainly in model rectangular geometries, along a tube bundle, or due to the perspective phenomenon in a limited region captured by a camera. While reviewing the state of the art, papers could not be found that address the subjects related to simultaneous CFD and PIV studies of fluid flow maldistribution in STHE on an industrial scale, with segmented baffles across a tube bundle and carried out by means of imaging of the complete cross-section the shell side. There were also no works concerned with identifying and analyzing the role of fluctuating vortex phenomena occurring in the bottom part of the shell side. Currently, designers and operators of STHE obtain information about the flow in these devices primarily from industry standards. Therefore, extending the state of knowledge with data obtained with the use of modern non-invasive measurement methods forms an important task from many points of view. Validation of the numerical model and assessment of the flow in the actual geometry of the cross-section of a tube bundle, taking into account the synergy of a variety of geometric and flow factors form a novelty in this research. Therefore, the purpose of the research was the use of CFD and PIV methods to demonstrate the role taken on by flow and geometrical parameters in the generation of variations of the local flow velocities in the device, which in turn lead to flow maldistribution and evolution of flow structures in the shell side. In addition, the results should serve to provide assistance to an extensive group of researchers who carry out CFD and PIV tests in flow systems containing tube bundles in a variety of devices.

2. Materials and Methods

2.1. Numerical Model and Computational Scheme

In numerical research, finite volume method (FVM) was utilized for the analysis of flow fields and flow parameters on the basis of Ansys Fluent CFD code. The pressure-based solution method segregated solver was applied for this purpose. The pressure-velocity coupling was performed with the semi-implicit method for pressure linked equations (SIMPLE) algorithm [28]. This algorithm was successfully utilized in research concerned with heat exchangers in a number of works [29–35]. Second-order upwind interpolation was used to determine representative samples of component values on the control volume surface and standard wall function. Variable values of under-relaxation factors were determined. They were equal to 0.6, 0.5, 0.8, and 0.8 for the pressure, momentum, turbulent kinetic energy, and turbulent energy dissipation, respectively. The following convergence criteria were adopted: 1×10^{-7} for the continuity equations, and 1×10^{-5} for others. Boundary conditions for the inlet: "velocity inlet" (turbulence intensity: 5%, hydraulic diameter: 0.1); for the outlet: "pressure outlet." A no slip boundary condition was used along the surface of the heat exchanger. Figure 1 presents the computational domain in an isometric view.

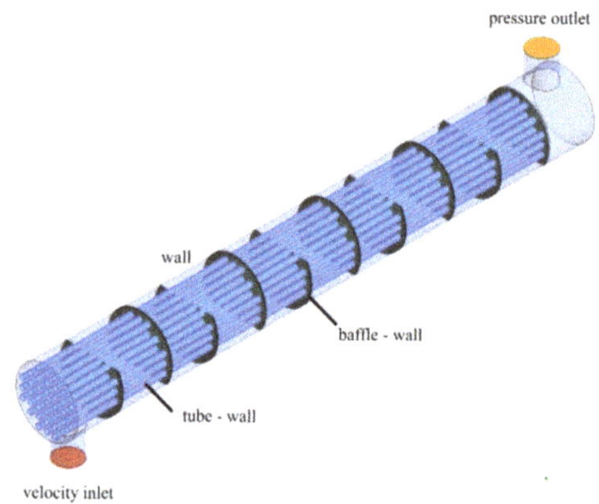

Figure 1. Isometric view of the computational domain.

The standard k-ε model was used as a closing model in simulations of turbulent flow incorporating Reynolds-averaged Navier–Stokes equations (RANS). It has been widely used for CFD research on heat exchangers [29–34,36]. The governing equations in the computational domain are presented in the following forms [37]:

continuity part:

$$\frac{\partial u_j}{\partial x_j} = 0 \tag{1}$$

momentum part:

$$\rho \frac{\partial \left(u_i u_j \right)}{\partial x_j} = -\frac{\partial p_i}{\partial x_i} + \frac{\partial}{\partial x_j}\left[\mu \left(\frac{\partial u_i}{\partial x_j} + \frac{\partial u_j}{\partial x_i} \right) \right] \tag{2}$$

energy part:

$$\rho \frac{\partial \left(u_j T \right)}{\partial x_j} = \frac{\partial}{\partial x_j}\left(\frac{\lambda}{C_p} \frac{\partial T}{\partial x_j} \right) \tag{3}$$

where u is the averaged velocity of the fluid [m/s]; p is pressure; ρ is the density of the fluid; T is temperature; μ is the kinematic viscosity of the fluid [m²/s]; C_p is the specific heat capacity [J/kgK]; and λ is the thermal conductivity [W/mK].

The regular k–e model is adopted to simulate turbulent flow in present paper, which is presented as:

turbulent kinetic energy part:

$$\rho \frac{\partial (k u_i)}{\partial x_i} = \frac{\partial}{\partial x_j}\left[\left(\mu + \frac{\mu_T}{\sigma_k} \right) \frac{\partial k}{\partial x_j} \right] + G_k - \rho \varepsilon \tag{4}$$

turbulent energy dissipation part:

$$\rho \frac{\partial (\varepsilon u_i)}{\partial x_i} = \frac{\partial}{\partial x_j}\left[\left(\mu + \frac{\mu_T}{\sigma_\varepsilon} \right) \frac{\partial \varepsilon}{\partial x_j} \right] + \frac{C_{1\varepsilon}\varepsilon}{k} G_k - C_{2\varepsilon} \rho \frac{\varepsilon^2}{k} \tag{5}$$

where k is turbulent kinetic energy [m²/s²], ε is turbulent dissipation rate [m²/s³], G_k is producing term of turbulent kinetic energy generated by mean velocity gradient, $C_{1\varepsilon}$ and $C_{2\varepsilon}$ are constants, σ_ε and σ_k are Prandtl numbers corresponding to turbulent kinetic energy and turbulent dissipation rate, and μ_T [Pas] is expressed as

$$\mu_T = \rho C_\mu \frac{k^2}{\varepsilon} \tag{6}$$

where $C_\mu = 0.09$ [-], $C_{1\varepsilon} = 1.44$ [-], $C_{2\varepsilon} = 1.92$ [-], $\sigma_k = 1.0$ [-], $\sigma_\varepsilon = 1.3$ [-], and G_k [-] is defined as

$$G_k = -\overline{\rho u'_i u'_j} \frac{\partial u_j}{\partial x_i} \tag{7}$$

2.2. Application of Particle Image Velocimetry Technique

The PIV technique forms one of the most common experimental methods that is applicable for verification of CFD models. It forms an optical method in which specific conditions of an experiment have to be fulfilled. It is necessary to ensure the transparency of the surface of the investigated device and illumination of the selected plane of the device with a coherent light, and thus the formation of a laser sheet plane. The illuminated phenomena are registered with a camera synchronized with the light source. In order to visualize the flow of the fluid, it is necessary to add suitably selected inert seeding particles to it in order to match the flow and physicochemical parameters of the fluid. The PIV technique uses the relationship between the displacement (in direction x and y defined as Δx and Δy respectively) of seeding particles recorded in the images and the known time between subsequent images often referred to as time between laser pulses Δt. Taking into account the above, the velocities u and v defined as Δx/Δt and Δy/Δt are calculated. On this basis, with the use of statistical methods, the velocity vector V which is expressed as $\sqrt{u^2 + v^2}$ can be determined (Figure 2).

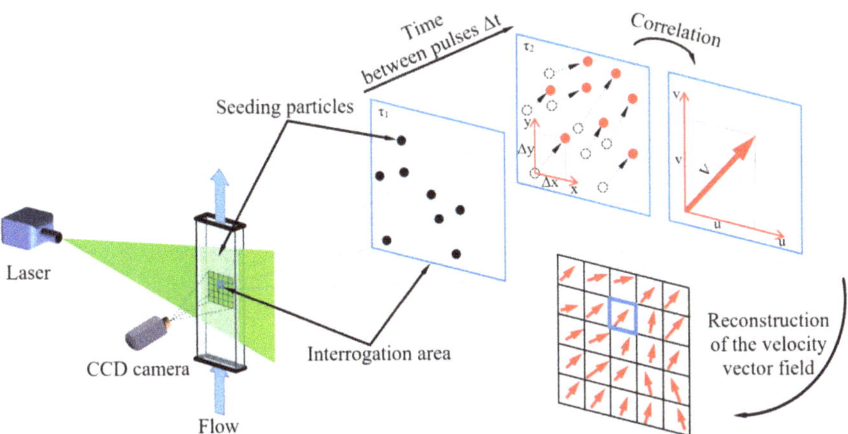

Figure 2. Idea of measurements applying the particle image velocimetry (PIV) technique.

In the research, the planar PIV method was used to visualize the fluid velocity field across a tube bundle. The model of the tested heat exchanger was made of Poly(methyl 2-methylpropenoate) (PMMA). Inside the exchanger, in the shell with an internal diameter D = 0.244 m, a tube bundle was installed, consisting of 37 tubes with an external diameter D_z = 0.02 m, mounted on 10 segmented baffles with a cut of 25%. Since the goal of the research involved the analysis of the phenomena in the shell side, the exchanger model did not include either head chambers or tubesheets; therefore, the fluid flow did not occur in the tube side (as the tubes were sealed on both sides). Water formed the fluid applied in the tests, and it was circulated in a sealed system. The liquid was extracted by a Grundfos

CR5-7 pump (Grundfos, Bjerringbro, Denmark) from a tank with the capacity of 0.2 m^3 and routed through a pipeline to the heat exchanger, and then returned to the tank. The pump was controlled by a signal from the ENKO EM-005C electromagnetic flowmeter (ENKO, Gliwice, Poland). The applied PIV method required that seeding particles were introduced into the flow. Fluorescent seeding particles PMMA-RHB-10 stained with rhodamine B were applied for this purpose. A stirrer was installed in the tank to maintain constant homogenization of the water. The type and number of seeding particles were determined on the basis of Adrian and Westerweel [38] and own research. The hydrodynamic characteristics of the heat exchanger model were evaluated for three liquid flow rates equal to 5 m^3/h (Re_{in} = 16,662), 7.5 m^3/h (Re_{in} = 24,993), and 10 m^3/h (Re_{in} = 33,324), respectively. These values corresponded to the inlet velocity v_{in} of 0.177 m/s, 0.266 m/s, and 0.354 m/s respectively. A Dantec Dynamics FlowSense EO-4M CCD camera (Dantec Dynamics, Skovlunde, Denmark) was installed in perpendicular plane to the cross-sectional plane of the shell side. The image was recorded at a resolution of 2048 × 2048 pixels in the double frame mode with a frequency of 10 Hz. The time between pulses was equal to 750, 500 and 300 μs for the flow rates of 5 m^3/h, 7.5 m^3/h, and 10 m^3/h, respectively. Two hundred double frame images were taken for each series of measurements. The measurement area was illuminated by a Dantec Dynamics DualPower TR Nd: YAG laser (Dantec Dynamics, Skovlunde, Denmark) via a laser sheet targeting mirror in such a way as to provide illumination of the imaging plane in a distance of 0.010 m from the last segmental baffle and in distance L_2 = 0.0135 m from the end of tube bundle. The justification for the selecting this location of the measurement plane is presented in Section 2.3. The optical system of the camera is equipped with an Omega Optical 550LP filter (Omega Optical Inc., Brattleboro, VT, USA), which cuts off more than 98% of the transmission of the light wave for wavelengths shorter from 5.5×10^{-7} m. As a result, no laser light with a wavelength of 5.32×10^{-7} m reaches the camera matrix, but only excited light emitted by fluorescent seeding particles, as it is characterized by the range of wavelength from 6.1×10^{-7} m to 6.5×10^{-7} m, i.e., above the cut-off threshold. The PIV calculations were performed using Dantec Dynamics Dynamic Studio software, ver. 2015a (Dantec Dynamics, Skovlunde, Denmark). The obtained test results are based on the analysis of the components of velocity u and v and the inlet velocity v_{in}. For the purposes of determining measure scale factor, a target located in the same plane as the generated laser sheet was used. During image calibration, no optical distortions were identified characteristic of imaging through curved surfaces or in the StereoPIV technique, where the cameras are inclined in relation to the measurement plane [38,39]. The use of images without optical correction, recorded in analogous optical paths, is common in the planar PIV technique [40–43]. Due to the assumed distance from the laser sheet and the size of the recorded image, the parallax error was not prevented. As a result, the field of view around the outer tubes in the tube bundle was excluded from analysis at a maximum distance of 1.07 mm (40 pixels) from the tube edge. Following the discussion in the work of Adrian and Westerweel [38], the uncertainty of estimating the displacement of seeding particles and the uncertainty of determining the velocity vectors have the greatest influence on an accuracy of the PIV measurement. Based on these two parameters, the maximum uncertainty of velocity determination in this experiment was set at 12%, which is illustrated by the corresponding error bars in Section 3.2.

The methodology of PIV calculations and analysis used during the research is presented in Figure 3. Figure 4 presents the scheme of the measuring station. Detailed dimensions of the model can be found in Figure 5 and Table 1.

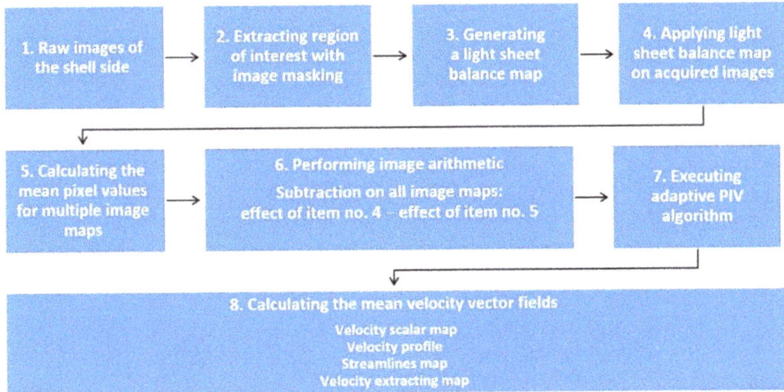

Figure 3. Applied PIV methodology.

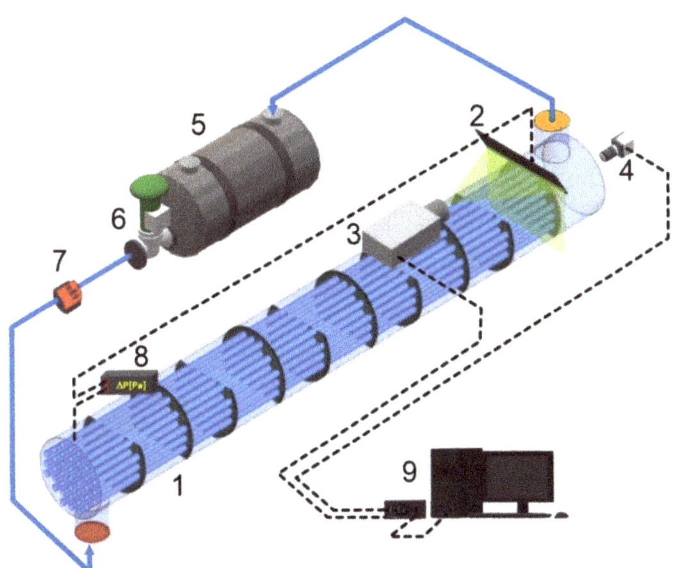

Figure 4. Experimental stand: shell-and-tube heat exchanger model (**1**), mirror (**2**), laser (**3**), CCD camera (**4**), tank (**5**), pump (**6**), flowmeter (**7**), differential pressure transmitter (**8**), and control and data acquisition station (**9**).

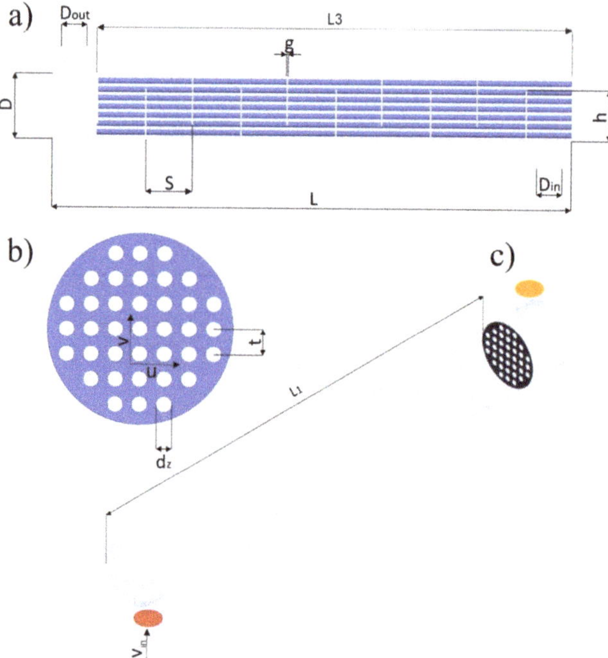

Figure 5. Geometry of the tested shell-and-tube heat exchanger (STHE) model. (**a**) Longitudinal section of the sell side; (**b**) cross-section of the shell side with the directions of velocities u and v; (**c**) location of the measuring area with direction of the velocity v_{in}.

Table 1. Detailed dimensions of the model.

Dimension	Value, m
D	0.244
D_{out}	0.1
D_{in}	0.1
S	0.178
g	0.0035
L	2
h	0.183
t	0.030
d_z	0.020
L1	1.7665
L2	0.0135
L3	1.8213

2.3. Assumptions and Conditions Applied in the Research

From the literature review we can note that the problem of perspective poses a challenge in the research applying tube bundles. The variable position of each tube in the bundle depending on its distance leads to the interferences in the visibility of the measurements (by obscuring the region of interest). This problem was already reported in previous works [22,25]. One of the solutions used in such conditions is to adjust an index of refraction of the fluid to the material from which the tubes are made. The results of the application of such approaches are presented in references [14,22]. In the current research, another solution was proposed, which involved the removal of the fraction of the tube bundle that obstructs the region of interest. CFD tests were carried out, and the results demonstrate that the removal of the tube section behind the final baffle only slightly affects the flow characteristics (Figure 6). The maximum differences of the dimensionless parameters defining the velocities differ by a maximum of 3.5%, which does not exceed the absolute differences between the results gained from simulations and ones gained on the basis of the experimental procedure. The pressure drops also vary within a small range that does not exceed 3%. Therefore, it was concluded that the geometric modifications performed in this research could provide a solution to the problem of perspective in visualizations across tube bundles.

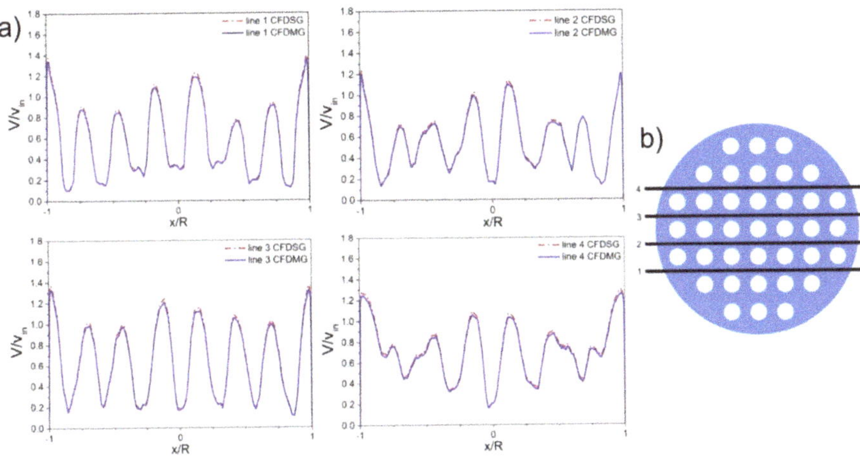

Figure 6. Comparison of velocity profiles for standard and modified tube bundle geometry. (**a**) Velocity profiles for Q = 5 m^3/h; (**b**) location of the lines 1–4.

3. Results and Discussion

3.1. Analysis of Sensitivity of Computational Mesh and Validation of Research Method

3.1.1. Analysis of Sensitivity of Computational Mesh

Before the discretization of the computational region was performed, the results were analyzed in terms of the optimization of the computational mesh. It was found that in the case of 3D heat exchanger models, tetrahedral type cells are most often employed in research. Therefore, inside the apparatus, a hybrid mesh with tetrahedral cells was utilized. Additionally, in order to maintain the correct density and proper values of the bridging functions in the shell side, a densified, five-layer mesh with hexahedral cells was used. In the next step, sensitivity analysis of the computational mesh density was performed. Four meshes with various densities were generated for this purpose (mesh 1: 8,435,989 elements; mesh 2: 10,266,557 elements; mesh 3: 12,112,572 elements; mesh 4: 12,972,099 elements). The layout of mesh cross-sections utilized in the research is presented in Figure 7. The reference for the sensitivity analysis applied the values of the recorded pressure drop in the heat exchanger (in relation to the results of experimental tests for the flow rate of 10 m^3/h) and the maximum velocities in the plane selected for the analysis. It was found that the mesh no. 3 offered considerable level of the consistency of the results combined with the lowest possible load of the computing resources (therefore, mesh 3 was included in further research). Additionally, by analyzing the obtained values of these parameters for the first three meshes, we can conclude that their values increase with the increase in the mesh density. In turn, for the fourth mesh with the greatest density, the results did not demonstrate a further increase in the values of the analyzed parameters. To illustrate this phenomenon, the obtained velocity distributions for individual meshes in the analyzed plane are presented in Figure 8. Additionally, Figure 9 shows the velocity distribution along an example of line 2 along the analyzed plane. A complete list of the analyzed numerical meshes coupled with the results is presented in Table 2.

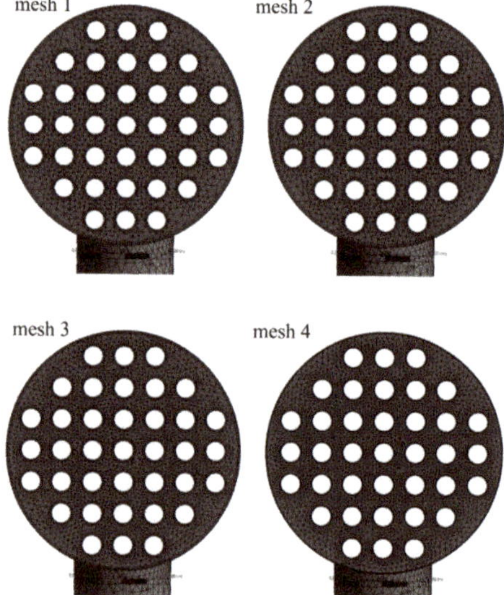

Figure 7. Layout of mesh cross-sections.

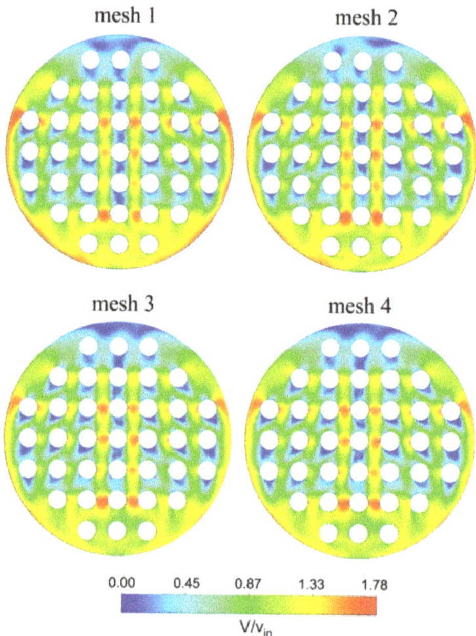

Figure 8. Velocity contours for analyzed geometric meshes.

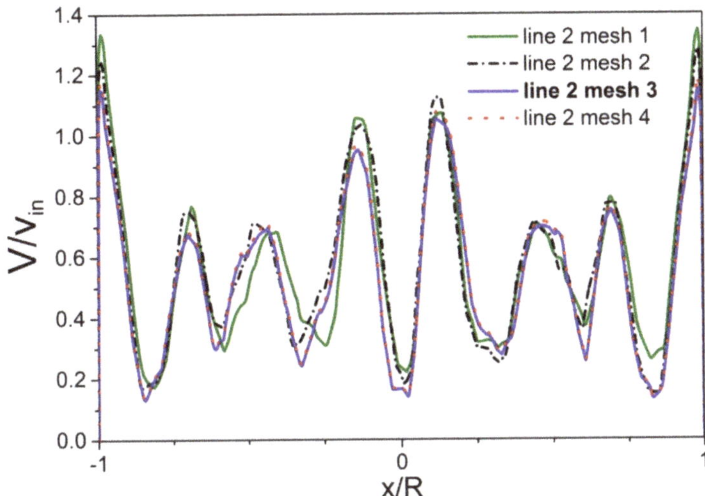

Figure 9. Velocity distributions for analyzed geometric meshes in line 2.

Table 2. Characteristics of computational fluid dynamics (CFD) meshes.

Parameter	Mesh 1	Mesh 2	Mesh 3	Mesh 4	Experimental Results and PIV
Total number of elements	8,435,989	10,266,557	12,112,572	12,972,099	—
Mean skewness	0.25	0.24	0.23	0.23	—
Maximum velocity [m/s]	0.68	0.67	0.65	0.65	0.58
Results compatibility [%]	85	87	89	89	—
Pressure drop ΔP [Pa]	2960	2930	2900	2904	2698
Results compatibility [%]	91	92	93	93	—

3.1.2. Validation of Research Methods

When several methods are used in research, it is necessary to validate the obtained results to ensure their credibility. For the purposes of this study, the results of investigations using CFD were validated by application of the results of experimental tests (in terms of pressure drop) and the PIV method (in terms of the velocity values obtained in the selected plane). The comparison was performed by application of three analyzed values of the volumetric flow rate (5 m³/h, 7.5 m³/h, and 10 m³/h). Table 3 presents the results of the comparative analysis. In the case of the pressure drop value, the calculation error was equal to, respectively, for Q = 5 m³/h—2%, for Q = 7.5 m³/h—2%, and for Q = 10 m³/h—7%. On the basis of the comparison of the second of the analyzed parameters (maximum velocity), the authors noted that the values of calculation errors were equal to: 10%, 10%, and 11%, respectively. In the case of the third of the considered parameters (mean velocity), the following calculation errors were noted: 6%, 8%, 9% for Q = 5 m³/h, 7.5 m³/h, and 10 m³/h, respectively.

Table 3. Validation of the results.

Parameter	Value		
Volumetric Flow rate [m^3/h]	5	7.5	10
Pressure drop (CFD) [Pa]	758	1693	2900
Pressure drop (exp. results) [Pa]	746	1656	2698
Results compatibility [%]	98	98	93
Maximum velocity CFD [m/s]	0.31	0.5	0.65
Maximum velocity PIV [m/s]	0.28	0.45	0.58
Results compatibility [%]	90	90	89
Mean velocity CFD [m/s]	0.17	0.26	0.35
Mean velocity PIV [m/s]	0.16	0.24	0.32
Results compatibility [%]	94	92	91

The obtained compatibility of both the pressure drop and the flow velocity is high and can form the basis for the statement that the adopted procedures, numerical models, and parameterization of the calculation conditions applied in the research process were adequate.

3.2. Analysis of a Liquid Flow Maldistribution across the Tube Bundle

One of the methods applicable for evaluating the maldistribution of the liquid flow through a tube bundle involves the analysis of velocity fields. This paper presents the results of the reconstruction of scalar velocity fields derived by both the PIV and CFD methods. On this basis, regions located in the cross-section of the shell side are identified, in which flow maldistribution occurs as a result of rapid variations in the hydrodynamic parameters of the liquid over time. The intensification of flow maldistribution in the shell side often leads to the development of adverse wake regions behind the tube. In these regions, the efficiency of the heat transfer process decreases. Flow visualization analysis can be applied for the purposes of locating zones that negatively affect the performance of heat exchangers, as well as those with a fully developed flow around the tubes. In Figures 10–12, in the central row of tubes in the bundle, we can observe typical zones with clearly reduced local velocity. They are located in the wake of each tube and have an orientation that is parallel to the direction of liquid flow. When an analysis is performed of the successive rows of tubes, and moving from the center of the shell, one can observe oblique deviations of the velocity reduction zones. This phenomenon is clearly symmetrical.

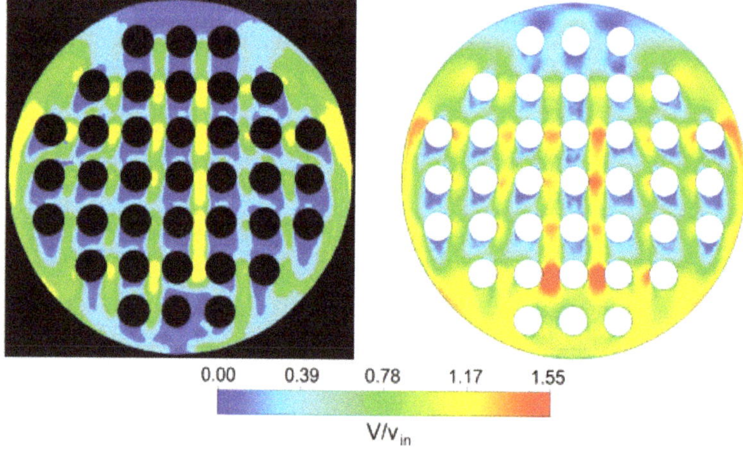

Figure 10. Comparison of scalar velocity fields V/v$_{in}$ for Q = 5m^3/h (CFD on the right, PIV on the left).

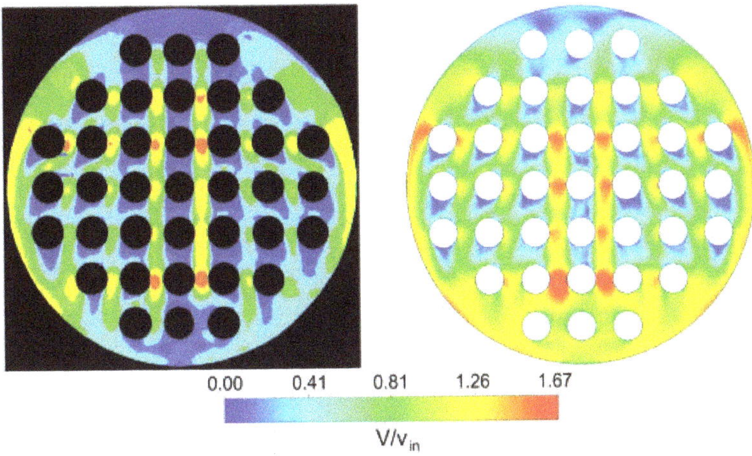

Figure 11. Comparison of scalar velocity fields V/v_{in} for $Q = 7.5\,m^3/h$ (CFD on the right, PIV on the left).

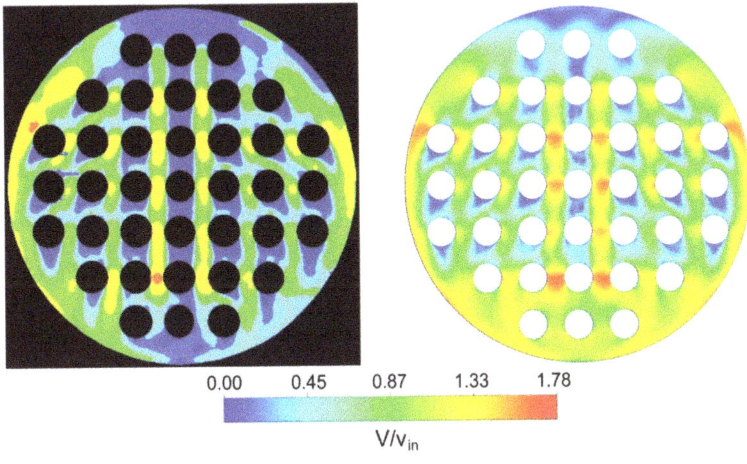

Figure 12. Comparison of scalar velocity fields V/v_{in} for $Q = 10\,m^3/h$ (CFD on the right, PIV on the left).

In addition, we can also note that the increase in the inflow rate of the liquid from $Q = 5\,m^3/h$ to $Q = 7.5\,m^3/h$ did not significantly affect the flow homogeneity or the distribution of the wake regions behind the tubes. However, a further increase in the flow rate to $Q = 10\,m^3/h$ resulted in the increase in velocity in these regions and their further reduction. The determination of the velocity profiles in the lines 1–4 marked in Figure 13 made it possible to quantify this observation (Figures 14–16). Thus, the variations in the velocity V/v_{in} in individual rows of tubes were compared for the flow rates of $Q = 7.5\,m^3/h$ and $Q = 10\,m^3/h$. Consequently, it was observed that the increase in the flow rate Q from $7.5\,m^3/h$ to $10\,m^3/h$ leads to a moderate increase in the mean velocity V/v_{in} by 20%, 10.42%, 6.25%, and 4.83% for the first, second, third, and fourth velocity profiles. Besides, on the basis of the comparison of the results obtained from PIV tests and calculations using CFD, we should note that the general flow diagrams overlap in most of the investigated cases. On the basis of the analysis of the velocity profiles in vertical lines (lines 5–8 in Figure 13), it was noted that the fluctuations of the liquid flow velocity may provide assistance in defining the start-up zone. In each of the analyzed cases, the fully developed flow specific for the middle part of the cross-section of the shell side, was developed only from the point when the first clear maximum of the velocity V/v_{in} was reached. In this region, there is a satisfactory compliance of the results gained by CFD and PIV methods. However, in the regions occupied by the first rows of tubes (both the horizontal and vertical velocity profiles) there were significant differences between the experimental and numerical results. It should be assumed that this is the effect of dynamic changes in this zone, where both the k-e model and the adopted PIV system parameters offer less accurate results. This is justified by the simplification applied in the calculation method, which does not take into account all the details of the multi-parameter flow evolution in this area, as well as the limitations of the experimental method, one drawback of which is associated with the decrease in the reliability in the regions characterized by considerable dynamic characteristics of velocity fluctuations. In regions characterized by smooth flow, the degree of conformity is satisfactory and does not exceed 10% of the maximum velocities. However, on the basis of the visual analysis of the flow, we can state that in the case of flow studies in the shell side, the CFD method provides results with smaller degree of detail in comparison to the application of PIV.

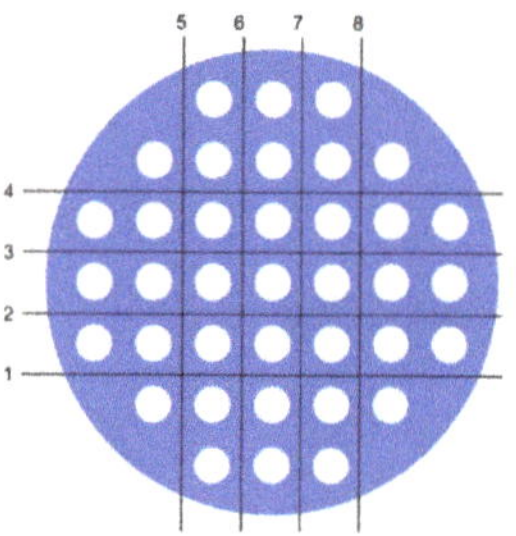

Figure 13. Location of lines applied for data extraction for purposes of determining velocity profiles.

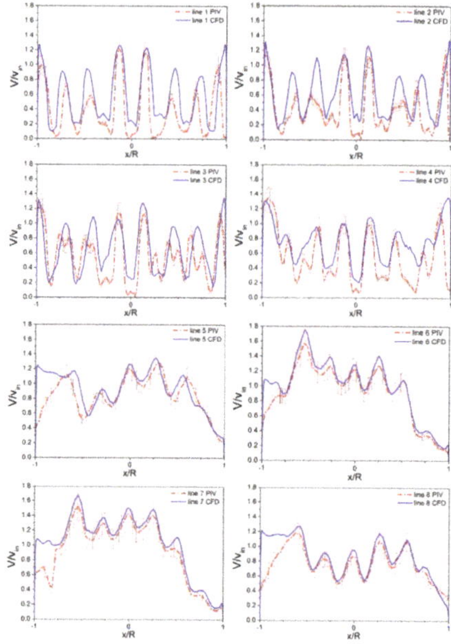

Figure 14. Comparison of velocity profiles derived by CFD method and gained from PIV for $Q = 5$ m^3/h.

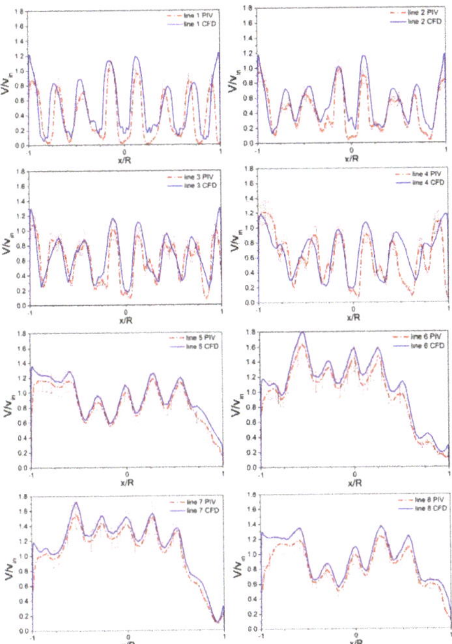

Figure 15. Comparison of velocity profiles derived by CFD method and gained from PIV for $Q = 7.5$ m^3/h.

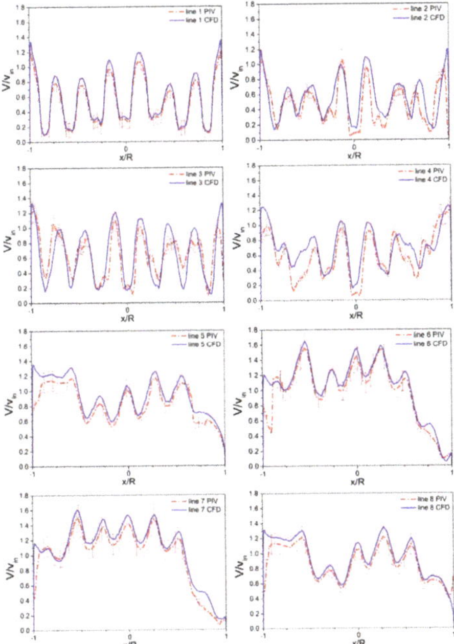

Figure 16. Comparison of velocity profiles derived by CFD method and gained from PIV for $Q = 10$ m^3/h.

The subsequent stage of the research into the maldistribution of the flow through shell side in STHE involved the analysis of the details on the direction of liquid flow. For this purpose, the vector velocity field and streamlines were applied. The enlarged region of the vector velocity field marked in Figure 17 indicates a satisfactory compliance of the results gained by the CFD and PIV methods. In both cases the converging regions and the diverging regions have been recognized, as well as quantitative velocity distribution along the successive rows of tubes was identified. The calculated results also demonstrate that the velocity reduction areas always form in the wake of the tubes, on the opposite side of the tubes in relation to the direction of the liquid inflow. Liquid streams flowing obliquely between the tubes are characterized by velocity that is clearly above the mean for the entire cross-section of the shell. It is one of the main factors responsible for the formation of swirl centers in the wake of the tubes. In these regions, the velocity also drops to values close to the minimum. However, due to the fact that an in-line tube bundle layout was applied in the research, the distance between oblique tubes in the consecutive rows is greater from the distance between the tubes in the same rows. As a result, velocity reduction areas do not extend to the subsequent tube rows (see also Figures 10–12). They are also characterized by a lower velocity gradient. Under such flow conditions ahead of the next tube, the velocity of the liquid increases, which can be considered a positive phenomenon.

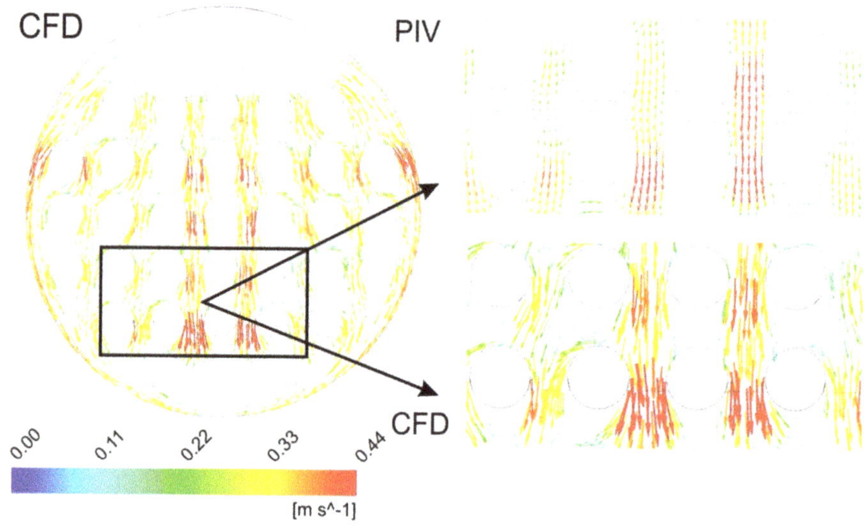

Figure 17. Sample comparison of vector velocity fields PIV and CFD for flow rate Q = 7.5 m^3/h.

Throughout the streamline analysis, both in the case of CFD and PIV method (Figure 18), the central streams, oblique streams and bypass streams were identified, and, very importantly, large vortex cells in the lower part of the shell side. The area occupied by the generated vortex comprised the lower surfaces of the last row of tubes and in particular the central tube. This vortex pattern is influenced by the central liquid stream and the two bypass streams. In the baffle window, these two types of streams tend to mix dynamically, and behind the baffle window, the direction of movement is additionally altered. As a consequence, extensive vortex patterns develop at the bottom of the shell. Due to the proportionality in the variations of the velocity of the central stream and bypass streams for the flow parameters investigated in the paper, we can conclude that the presence of vortex patterns in the baffle window is not dependent on the inflow rate Q. Therefore, we should forecast that the factor responsible for the intensity and characteristics the vortex zone in the baffle window is associated with the geometric parameters. The streamlines analysis indicates considerable conformity of the CFD and PIV results. Only the area between the shell and the first row of tubes is characterized by significant differences in the streamlines. These results are also reflected in the distribution of the vertical velocity profiles (lines 5–8 in Figures 14–16). In order to obtain knowledge about the dynamics of large vortex cells in the lower part of the shell side, the variability of the velocity of liquid flow in this area was examined in the further part of the study.

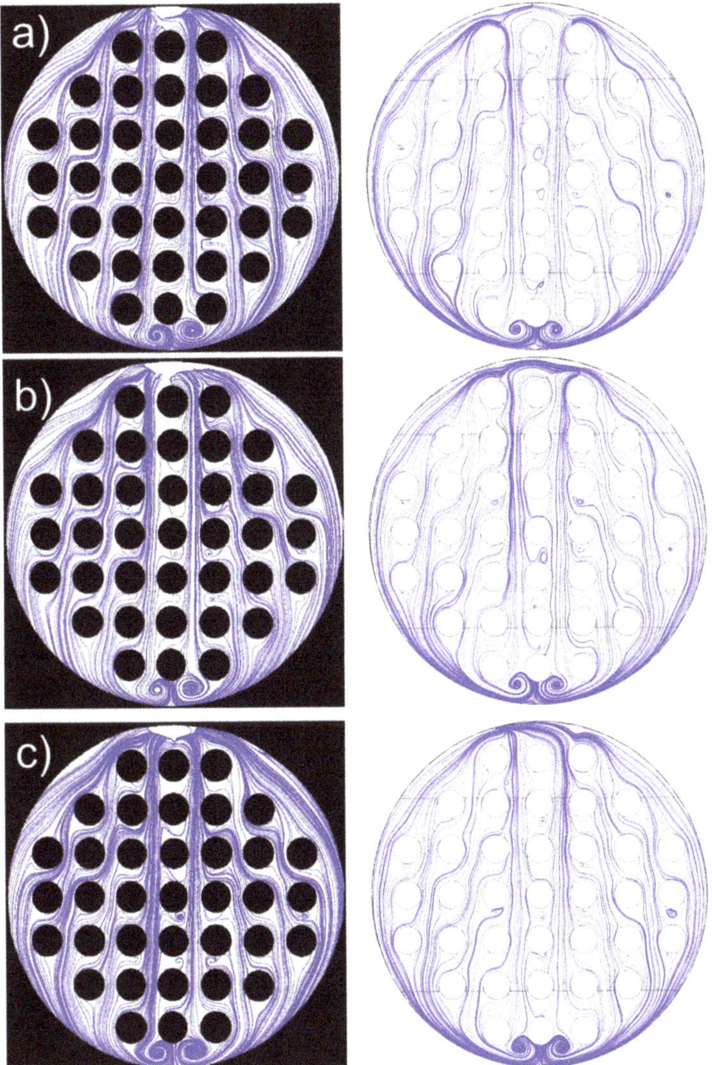

Figure 18. Comparison of streamlines PIV (left row) and CFD (right row) for various liquid flow rate: (**a**) Q = 5 m^3/h, (**b**) Q = 7.5 m^3/h, and (**c**) Q = 10 m^3/h.

The occurrence of displacement of vortex patterns is also noteworthy. This is illustrated in Figure 19 by determining the variations in the velocity in time at point F, located in the middle of the distance between the final tube in the middle row and on the shell wall of the heat exchanger. Taking into account the theoretical velocity distribution in the vortex structure, characterized by an increase in velocity on the borders of the vortex and a decrease in velocity in its core, it can be demonstrated that in the area of the border rows of the tubes, we have to do with recurrent fluctuations in the vortex location. Regardless of the flow rate Q, these fluctuations are at a similar level (see dynamic velocity V/v_{in}

changes in Figure 19). Thus, it is possible that the observed mechanism of the vortex displacement may generate erosive interactions on the outer surfaces of tubes in such regions, occurring by analogy to the erosive interactions inside the tubes, in tube bends [44]. An additional factor that supports such observation is associated with the fact that contaminants accumulate on the bottom of the STHE shell, which may enhance the erosive effect of the vortex structures. This area should therefore be monitored for accelerated erosive wear. In extreme conditions, the fluctuation of vortices in the lower part of the shell side could interfere in the symmetry of the flow and contribute to imbalanced heat transfer. The solution that could counteract this situation could be associated with geometrical modifications in the form of bars preventing the displacement of vortices.

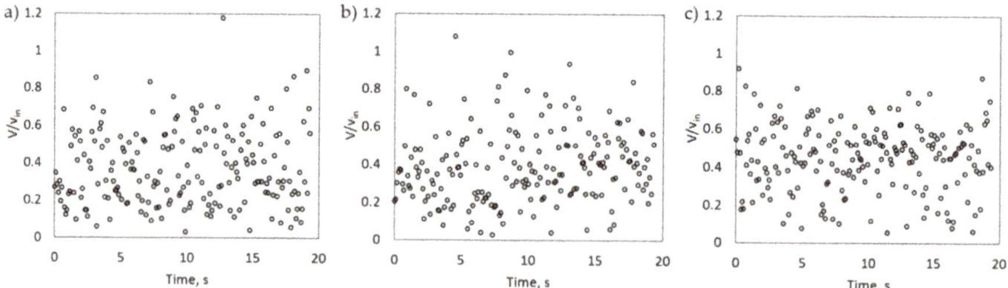

Figure 19. Variations in velocity for point F in time for (**a**) $Q = 5$ m^3/h, (**b**) $Q = 7.5$ m^3/h, and (**c**) $Q = 10$ m^3/h.

4. Conclusions

On the basis of the numerical and experimental investigations of the shell side in STHEs, it was found that the information in the images of the liquid velocity fields is sufficient for the purposes of the effective evaluation of flow maldistribution. On the basis of the analysis, the following conclusions were formulated:

1. The study presented a method of eliminating perspective phenomenon on the recorded image of the shell side resulting from the geometric modification involving the removal of a section of the tube bundle behind the last baffle. This procedure extends the area of the image feasible for the planar PIV analysis of the full cross-section resulting from the actual geometry of the shell side. At the same time, the variations in the hydrodynamic parameters resulting from the removal of a section of the tube bundle remain at an acceptable level. The maximum differences of the dimensionless parameters defining the velocities differ by a maximum of 3.5%, also pressure drops vary within a small range that does not exceed 3%.
2. The results obtained in the validation process (in relation to the values of pressure drops obtained on the basis of experimental tests and the values of average and maximum velocities - in relation to the PIV tests) of the adopted numerical strategy allow us to conclude that, in case of the numerical calculations in the shell side of a STHE, the use of hybrid mesh with tetrahedral cells and the standard k-ε model with RANS enables reliable analysis of the flow inside the heat exchanger.
3. Liquid flow maldistributions were identified in the studies. Two central streams could be distinguished, flowing in parallel on both sides of the middle row of tubes. Between the tubes in the middle row, regardless of the flow rate Q, there were cyclic areas with reduced velocity and high vortex generation potential. Their occurrence was considered unfavorable due to their negative role in forming heat transfer parameters.
4. In the areas on the outer sides of the central streams there were symmetrical oblique streams, where the regions of reduced velocities did not extend to the subsequent tube rows. Oblique liquid flow in an in-line layout transforms the flow patterns toward structures specific to liquid flow into a staggered layout.

5. An increase in the velocity gradient was observed with an increase in the flow rate Q for the successive tube rows. The highest velocity gradient was noted for line 1 and the lowest for line 4 located in Figure 13. On this basis, it was concluded that the evolution of velocity in the tested geometry is an important factor that should be taken into account in thermal calculations.

6. The phenomenon of the displacement of vortices in the baffle window was observed. This can potentially lead to unfavorable erosive effects on tubes within the vortex zone. Structural elements in the form of bars in the lower part of the exchanger shell may prevent the displacement of vortices in the baffle window. Such modifications are likely to extend the effective service life of the heat exchanger.

Author Contributions: Conceptualization, G.L.; methodology, M.W. and S.K.; software, M.W. and G.L.; validation, D.Z., M.W. and S.K.; formal analysis, M.W. and G.L.; investigation, G.L. and M.W.; resources, S.K.; data curation, S.K.; writing—Original draft preparation, G.L. and M.W.; writing—Review and editing, D.Z.; visualization, G.L. and M.W.; supervision, G.L.; project administration, S.K. and D.Z.; funding acquisition, D.Z. All authors have read and agreed to the published version of the manuscript.

Funding: This research was carried out with the support of the Interdisciplinary Centre for Mathematical and Computational Modelling (ICM) University of Warsaw under grant no G73-10.

Conflicts of Interest: The authors declare no conflict of interest

References

1. Bhutta, M.M.A.; Hayat, N.; Bashir, M.H.; Khan, A.R.; Ahmad, K.N.; Khan, S. CFD applications in various heat exchangers design: A review. *Appl. Therm. Eng.* **2012**, *32*, 1–12. [CrossRef]

2. Zhang, Z.; Li, Y. CFD simulation on inlet configuration of plate-fin heat exchangers. *Cryogenics* **2003**, *43*, 673–678. [CrossRef]

3. Yataghene, M.; Pruvost, J.; Fayolle, F.; Legrand, J. CFD analysis of the flow pattern and local shear rate in a scraped surface heat exchanger. *Chem. Eng. Process. Process. Intensif.* **2008**, *47*, 1550–1561. [CrossRef]

4. Zhang, L.-Z. Flow maldistribution and thermal performance deterioration in a cross-flow air to air heat exchanger with plate-fin cores. *Int. J. Heat Mass Transf.* **2009**, *52*, 4500–4509. [CrossRef]

5. Kim, M.I.; Lee, Y.; Kim, B.W.; Lee, D.H.; Song, W.S. CFD modeling of shell-and-tube heat exchanger header for uniform distribution among tubes. *Korean J. Chem. Eng.* **2009**, *26*, 359–363. [CrossRef]

6. Ozden, E.; Tari, I. Shell side CFD analysis of a small shell-and-tube heat exchanger. *Energy Convers. Manag.* **2010**, *51*, 1004–1014. [CrossRef]

7. Wang, Q.Y.; Dong, M.L. Characteristics of fluid flow and heat transfer in shell side of heat exchangers with longitudinal flow of shell side fluid with different supporting structures. In Proceedings of the International Conference on Power Engineering, Singapore, 3–6 December 2007.

8. Wang, Q.; Chen, Q.; Chen, G.; Zeng, M. Numerical investigation on combined multiple shell-pass shell-and-tube heat exchanger with continuous helical baffles. *Int. J. Heat Mass Transf.* **2009**, *52*, 1214–1222. [CrossRef]

9. Mohammadi, K.; Heidemann, W.; Muller-Steinhagen, H. Numerical investigation of the effect of baffle orientation on heat transfer and pressure drop in a shell and tube heat exchanger with leakage flows. *Heat Transf. Eng.* **2009**, *30*, 1123–1135. [CrossRef]

10. Raj, V.A.A.; Velraj, R. Heat transfer and pressure drop studies on a PCM-heat exchanger module for free cooling applications. *Int. J. Therm. Sci.* **2011**, *50*, 1573–1582. [CrossRef]

11. Sun, Y.; Wang, X.; Long, R.; Yuan, F.; Yang, K. Numerical investigation and optimization on shell side performance of a shell and tube heat exchanger with inclined trefoil-hole baffles. *Energies* **2019**, *12*, 4138. [CrossRef]

12. Jun, S.; Puri, V.M. 3D milk-fouling model of plate heat exchangers using computational fluid dynamics. *Int. J. Dairy Technol.* **2005**, *58*, 214–224. [CrossRef]

13. de Bonis, M.V.; Ruocco, G. Conjugate fluid flow and kinetics modeling for heat exchanger fouling simulation. *Int. J. Therm. Sci.* **2009**, *48*, 2006–2012. [CrossRef]

14. Iwaki, C.; Cheong, K.-H.; Monji, H.; Matsui, G. PIV measurement of the vertical cross-flow structure over tube bundles. *Exp. Fluids* **2004**, *37*, 350–363. [CrossRef]

15. Iwaki, C.; Cheong, K.-H.; Monji, H.; Matsui, G. Vertical, bubbly, cross-flow characteristics over tube bundles. *Exp. Fluids* **2005**, *39*, 1024–1039. [CrossRef]

16. Zhang, X.; Daghrah, M.; Wang, Z.; Liu, Q.; Jarman, P.; Negro, M. Experimental verification of dimensional analysis results on flow distribution and pressure drop for disc-type windings in OD cooling modes. *IEEE Trans. Power Deliv.* **2017**, *33*, 1647–1656. [CrossRef]

17. Talapatra, S.; Farrell, K. Planar PIV experiments inside a transparent shell-and-tube exchanger. *SME Int. Mech. Eng. Congr. Expo. Proc.* **2014**, *7*. [CrossRef]

18. Bouhairie, S.; Talapatra, S.; Farrell, K. Turbulent flow in a no-tube-in-window shell-and-tube heat exchanger: CFD vs PIV. *ASME Int. Mech. Eng. Congr. Expo. Proc.* **2014**, *7*. [CrossRef]

19. Talapatra, S.; Farrell, K. Two-phase flow regimes in exchangers and piping: Part 1. *ASME Int. Mech. Eng. Congr. Expo. Proc* **2015**, *57465*. [CrossRef]

20. Wang, K.; Zhang, Z.; Liu, Q.; Tu, X.; Kim, H.-B. PIV measurement of tube-side in a shell and tube heat exchanger. *Pol. J. Chem. Technol.* **2018**, *20*, 60–66. [CrossRef]

21. Liu, P.; Zheng, N.; Rui, L.; Shan, F.; Liu, Z.; Liu, W. PIV measurement of flow structures in a circular heat exchange tube with central slant rod inserts. *Energy Procedia* **2017**, *142*, 3793–3798. [CrossRef]

22. Lee, S.; Delgado, M.; Hassan, Y.; Lee, S.J. Experimental investigation of the isothermal flow field across slant 5-tube bundles in helically coiled steam generator geometry using PIV. *Nucl. Eng. Des.* **2018**, *338*, 261–268. [CrossRef]

23. Delgado, M.; Lee, S.; Hassan, Y.; Anand, N.K. Flow visualization study at the interface of alternating pitch tube bundles in a model helical coil steam generator using particle image velocimetry. *Int. J. Heat Mass Transf.* **2018**, *122*, 614–628. [CrossRef]

24. Delgado, M.; Hassan, Y.; Anand, N.K. Experimental flow visualization study using particle image velocimetry in a helical coil steam generator with changing lateral pitch geometry. *Int. J. Heat Mass Transf.* **2019**, *133*, 756–768. [CrossRef]

25. Im, S.; Kim, H.T.; Rhee, B.W.; Sung, H.J. PIV measurements of the flow patterns in a CANDU-6 model. *Ann. Nucl. Energy* **2016**, *98*, 1–11. [CrossRef]

26. Chang, T.-H.; Lee, C.-H.; Lee, H.-S.; Lee, K.-S. Velocity profiles between two baffles in a shell and tube heat exchanger. *J. Therm. Sci.* **2015**, *24*, 356–363. [CrossRef]

27. Wen, J.; Yang, H.; Wang, S.; Gu, X. PIV experimental investigation on shell-side flow patterns of shell and tube heat exchanger with different helical baffles. *Int. J. Heat Mass Transf.* **2017**, *104*, 247–259. [CrossRef]

28. Fluent Inc. In *FLUENT 6.3 User s Guide. Options*; Fluent Inc. Centerra Resource Park 10 Cavendish Court: Lebanon, NH, USA, 2006.

29. Yang, J.; Liu, W. Numerical investigation on a novel shell-and-tube heat exchanger with plate baffles and experimental validation. *Energy Convers. Manag.* **2015**, *101*, 689–696. [CrossRef]

30. Shahril, S.; Quadir, G.; Amin, N.; Badruddin, I.A. Thermo hydraulic performance analysis of a shell-and-double concentric tube heat exchanger using CFD. *Int. J. Heat Mass Transf.* **2017**, *105*, 781–798. [CrossRef]

31. el Maakoul, A.; Laknizi, A.; Saadeddine, S.; el Metoui, M.; Zaite, A.; Meziane, M.; Ben-Abdellah, A. Numerical comparison of shell-side performance for shell and tube heat exchangers with trefoil-hole, helical and segmental baffles. *Appl. Therm. Eng.* **2016**, *109*, 175–185. [CrossRef]

32. Pal, E.; Kumar, I.; Joshi, J.; Maheshwari, N. CFD simulations of shell-side flow in a shell-and-tube type heat exchanger with and without baffles. *Chem. Eng. Sci.* **2016**, *143*, 314–340. [CrossRef]

33. You, Y.; Chen, Y.; Xie, M.; Luo, X.; Jiao, L.; Huang, S. Numerical simulation and performance improvement for a small size shell-and-tube heat exchanger with trefoil-hole baffles. *Appl. Therm. Eng.* **2015**, *89*, 220–228. [CrossRef]

34. Leoni, G.B.; Klein, T.S.; Medronho, R.D.A. Assessment with computational fluid dynamics of the effects of baffle clearances on the shell side flow in a shell and tube heat exchanger. *Appl. Therm. Eng.* **2017**, *112*, 497–506. [CrossRef]

35. Nogueira, X.; Taylor, B.J.; Gomez, H.; Colominas, I.; Mackley, M.R. Experimental and computational modeling of oscillatory flow within a baffled tube containing periodic-tri-orifice baffle geometries. *Comput. Chem. Eng.* **2013**, *49*, 1–17. [CrossRef]

36. Ambekar, A.S.; Sivakumar, R.; Anantharaman, N.; Vivekenandan, M. CFD simulation study of shell and tube heat exchangers with different baffle segment configurations. *Appl. Therm. Eng.* **2016**, *108*, 999–1007. [CrossRef]
37. Versteeg, H.K.; Malalasekera, W.; Orsi, G.; Ferziger, J.H.; Date, A.W.; Anderson, J.D. *An Introduction to Computational Fluid Dynamics—The Finite Volume Method*; Pearson Education Limited: London, UK, 1995; ISBN 9783540594345.
38. Adrian, R.J.; Westerweel, J. *Particle Image Velocimetry*; Cambridge University Press: New York, NY, USA, 2011; ISBN 978-0-521-44008-0.
39. Tropea, C.; Yarin, A.L.; Foss, J.F. *Springer Handbook of Experimental Fluid Mechanics*; Springer: Berlin/Heidelberg, Germany, 2007.
40. Binyet, E.M.; Chang, J.-Y.; Huang, C.-Y. Flexible plate in the wake of a square cylinder for piezoelectric energy harvesting—Parametric study using fluid–structure interaction modeling. *Energies* **2020**, *13*, 2645. [CrossRef]
41. Matsubara, T.; Shima, Y.; Aono, H.; Ishikawa, H.; Segawa, T. Effects of jet induced by string-type plasma actuator on flow around three-dimensional bluff body and drag force. *Energies* **2020**, *13*, 872. [CrossRef]
42. Liu, D.; Wang, Y.Z.; Shi, W.; Kim, H.-B.; Tang, A. Slit wall and heat transfer effect on the taylor vortex Flow. *Energies* **2015**, *8*, 1958–1974. [CrossRef]
43. Paul, S.; Tachie, M.F.; Ormiston, S. Experimental study of turbulent cross-flow in a staggered tube bundle using particle image velocimetry. *Int. J. Heat Fluid Flow* **2007**, *28*, 441–453. [CrossRef]
44. Gogolin, A.; Wasilewski, M.; Ligus, G.; Wojciechowski, S.; Gapiński, B.; Krolczyk, J.B.; Zajac, D.; Krolczyk, G. Influence of geometry and surface morphology of the U-tube on the fluid flow in the range of various velocities. *Measurement* **2020**, *164*, 108094. [CrossRef]

MDPI
St. Alban-Anlage 66
4052 Basel
Switzerland
Tel. +41 61 683 77 34
Fax +41 61 302 89 18
www.mdpi.com

Energies Editorial Office
E-mail: energies@mdpi.com
www.mdpi.com/journal/energies